Biofabrication
Micro- and Nano-Fabrication, Printing, Patterning, and Assemblies

Biofabrication
Micro- and Nano-Fabrication, Printing, Patterning, and Assemblies

Edited by

Gabor Forgacs

Wei Sun

AMSTERDAM • BOSTON • HEIDELBERG • LONDON
NEW YORK • OXFORD • PARIS • SAN DIEGO
SAN FRANCISCO • SINGAPORE • SYDNEY • TOKYO

William Andrew is an imprint of Elsevier

William Andrew is an imprint of Elsevier
The Boulevard, Langford Lane, Kidlington, Oxford OX5 1GB, UK
225 Wyman Street, Waltham, MA 02451, USA

Notice
No responsibility is assumed by the publisher for any injury and/or damage to persons
or property as a matter of products liability, negligence or otherwise, or from any use or
operation of any methods, products, instructions or ideas contained in the material herein.
Because of rapid advances in the medical sciences, in particular, independent verification
of diagnoses and drug dosages should be made.

British Library Cataloguing-in-Publication Data
A catalogue record for this book is available from the British Library

Library of Congress Cataloging-in-Publication Data
A catalog record for this book is available from the Library of Congress

ISBN: 978-1-4557-2852-7

For information on all William Andrew publications
visit our web site at books.elsevier.com

Typeset by MPS Limited, Chennai, India
www.adi-mps.com

Printed and bound by CPI Group (UK) Ltd, Croydon, CR0 4YY

12 13 14 15 16   9 8 7 6 5 4 3 2 1

Working together
to grow libraries in
developing countries

www.elsevier.com • www.bookaid.org

Contents

Preface

Biofabrication, as the term reveals, combines the biological sciences with the physical sciences and/or engineering approaches, more generally, with technologies that assemble biologically active or biologically derived systems. Biofabrication generates models, systems, devices, interfaces, and products that can be used for a wide range of applications in tissue science and tissue engineering, disease pathogeneses studies, and the development of new drugs. One distinguishing characteristics of biofabrication is that although it relies on input from physics, chemistry, and engineering, it provides output that often triggers further developments in these disciplines.

This interplay can be well illustrated with applications in regenerative medicine, especially in the latest efforts to mitigate the chronic and increasing problem of donor organ shortage. The goal of tissue engineering is the replacement of diseased, damaged, or missing tissues or organs by growing cells—traditionally in artificial scaffolds, which are convenient extracellular matrix mimics—until functional tissues are created. Biofabrication goes beyond tissue engineering in that it can develop the optimal "hardware" to build the organ and tissue substitutes. In particular, biofabrication methods led to the adaptation of three-dimensional additive manufacturing technologies—such as 3D printing—for building architecturally complex tissue scaffolds and biological structures. In turn, 3D bioprinting led to novel engineering solutions that aided in the development of new tissue engineering methods, such as scaffold-free tissue engineering, which relies exclusively on the inherent self-assembly properties of cells and tissues.

Another example of this interplay is the role of biofabrication in the integration of biological components with electronics. This assisted in the development of biosensors—devices that couple the molecular recognition capabilities of biological systems with the data processing capabilities of electronics for rapid, sensitive, and selective analysis.

This book, the first of its kind, introduces the most salient approaches and methods used by practitioners of this emerging discipline. The authors are leaders in the field, with many of them pioneers of new technologies or solutions to outstanding problems at the interface of the medical, life, physical, and engineering sciences. We hope readers will share our enthusiasm about biofabrication and that this book will encourage them to explore this ultimately multidisciplinary field.

Finally, the editors wish to acknowledge the support and dedication of the many colleagues and friends in the field of biofabrication, as well as the support of the International Society of Biofabrication and the journal *Biofabrication*, which made the publication of this book possible.

Gabor Forgacs
Wei Sun

List of Contributors

Muhammad Ali
INSERM U1026, Bordeaux, F-33076 France; Université Bordeaux Segalen, Bordeaux, F-33076 France

William E. Bentley
Institute for Bioscience and Biotechnology Research, University of Maryland, College Park, MD, USA; Fischell Department of Bioengineering, University of Maryland, College Park, MD, USA

Andrew M. Blakely
Department of Molecular Pharmacology, Physiology, and Biotechnology, Center for Biomedical Engineering, Brown University, Providence, RI, USA

Gulden Camci-Unal
Center for Biomedical Engineering, Department of Medicine, Brigham and Women's Hospital, Harvard Medical School, Cambridge, MA, USA; Harvard-MIT Division of Health Sciences and Technology, Massachusetts Institute of Technology Cambridge, MA, USA

Sylvain Catros
INSERM U1026, Bordeaux, F-33076 France; Université Bordeaux Segalen, Bordeaux, F-33076 France

Peter R. Chai
Department of Molecular Pharmacology, Physiology, and Biotechnology, Center for Biomedical Engineering, Brown University, Providence, RI, USA

M. Dean Chamberlain
Department of Chemical Engineering and Applied Chemistry, Institute of Biomaterials and Biomedical Engineering, University of Toronto, Toronto, Canada

U Kei Cheang
Drexel University, Department of Mechanical Engineering & Mechanics, Philadelphia, PA, USA

Shaochen Chen
Department of NanoEngineering, University of California, San Diego, La Jolla, CA, USA

Yi Cheng
Department of Materials Science and Engineering, University of Maryland, College Park, MD, USA; Institute for Systems Research, University of Maryland, College Park, MD, USA

Mohit P. Chhaya
Institute of Health and Biomedical Innovation, Queensland University of
Technology, Brisbane, Australia

Dong-Woo Cho
Department of Mechanical Engineering, POSTECH, South Korea. Division of
Integrative Biosciences and Biotechnology, POSTECH, South Korea

Ema C. Ciucurel
Department of Chemical Engineering and Applied Chemistry, Institute
of Biomaterials and Biomedical Engineering, University of Toronto, Toronto,
Canada

Loredana De Bartolo
Institute on Membrane Technology, National Research Council of Italy, ITM-CNR,
c/o University of Calabria, via P. Bucci cubo 17/C, I-87030 Rende (CS), Italy

Alexandre Ducom
INSERM U1026, Bordeaux, F-33076 France; Université Bordeaux Segalen,
Bordeaux, F-33076 France

Jean-Christophe Fricain
INSERM U1026, Bordeaux, F-33076 France; Université Bordeaux Segalen,
Bordeaux, F-33076 France

Reza Ghodssi
Institute for Systems Research, University of Maryland, College Park, MD,
USA; Department of Electrical and Computer Engineering, University of
Maryland, College Park, MD, USA

Fabien Guillemot
INSERM U1026, Bordeaux, F-33076 France; Université Bordeaux Segalen,
Bordeaux, F-33076 France

Bertrand Guillotin
INSERM U1026, Bordeaux, F-33076 France; Université Bordeaux Segalen,
Bordeaux, F-33076 France

Qudus Hamid
Department of Mechanical Engineering and Mechanics, Drexel University,
Philadelphia, PA, USA

Young Hun Jeong
Department of Mechanical Engineering, Korea Polytechnic University,
South Korea

Dietmar W. Hutmacher
Institute of Health and Biomedical Innovation, Queensland University of
Technology, Brisbane, Australia; George W. Woodruff School of Mechanical
Engineering, Georgia Institute of Technology, Atlanta, GA, USA

Min Jun Kim
School of Biomedical Engineering, Science & Health Systems, Drexel
University, Philadelphia, PA, USA

Virginie Keriquel
INSERM U1026, Bordeaux, F-33076 France; Université Bordeaux Segalen,
Bordeaux, F-33076 France

Ali Khademhosseini
Center for Biomedical Engineering, Department of Medicine, Brigham
and Women's Hospital, Harvard Medical School, Cambridge, MA, USA;
Harvard-MIT Division of Health Sciences and Technology, Massachusetts
Institute of Technology Cambridge, MA, USA; Wyss Institute for Biologically
Inspired Engineering, Harvard University, Boston, MA, USA

Joydip Kundu
Department of Mechanical Engineering, POSTECH, South Korea

Benjamin D. Liba
Institute for Bioscience and Biotechnology Research, University of Maryland,
College Park, MD, USA

Yi Liu
Institute for Bioscience and Biotechnology Research, University of Maryland,
College Park, MD, USA

Ferry P.W. Melchels
Institute of Health and Biomedical Innovation, Queensland University of
Technology, Brisbane, Australia; Department of Orthopaedics, University
Medical Center, Utrecht, the Netherlands

Antonietta Messina
Department of Chemical Engineering and Materials University of Calabria,
via P. Bucci cubo 45/A, I-87030 Rende (CS), Italy; Institute on Membrane
Technology, National Research Council of Italy, ITM-CNR, c/o University
of Calabria, via P. Bucci cubo 17/C, I-87030 Rende (CS), Italy

Jeffrey R. Morgan
Department of Molecular Pharmacology, Physiology, and Biotechnology, Center
for Biomedical Engineering, Brown University, Providence, RI, USA

Koichi Nakayama
Graduate School of Science and Engineering, Saga University, Japan

Anthony P. Napolitano
Department of Molecular Pharmacology, Physiology, and Biotechnology, Center for Biomedical Engineering, Brown University, Providence, RI, USA

Falguni Pati
Department of Mechanical Engineering, POSTECH, South Korea

Gregory F. Payne
Institute for Bioscience and Biotechnology Research, University of Maryland, College Park, MD, USA; Fischell Department of Bioengineering, University of Maryland, College Park, MD, USA

Adam P. Rago
Department of Molecular Pharmacology, Physiology, and Biotechnology, Center for Biomedical Engineering, Brown University, Providence, RI, USA

Murielle Remy
INSERM U1026, Bordeaux, F-33076 France; Université Bordeaux Segalen, Bordeaux, F-33076 France

Gary W. Rubloff
Department of Materials Science and Engineering, University of Maryland, College Park, MD, USA; Institute for Systems Research, University of Maryland, College Park, MD, USA

Jan T. Schantz
Department of Plastic and Hand Surgery, Klinikum Rechts der Isar, Technical University Munich, Germany

Jacquelyn Y. Schell
Department of Molecular Pharmacology, Physiology, and Biotechnology, Center for Biomedical Engineering, Brown University, Providence, RI, USA

Michael V. Sefton
Department of Chemical Engineering and Applied Chemistry, Institute of Biomaterials and Biomedical Engineering, University of Toronto, Toronto, Canada

Jessica Snyder
Department of Mechanical Engineering and Mechanics, Drexel University, Philadelphia, PA, USA

Pranav Soman
Department of NanoEngineering, University of California, San Diego, La Jolla, CA, USA

Agnès Souquet
INSERM U1026, Bordeaux, F-33076 France; Université Bordeaux Segalen, Bordeaux, F-33076 France

Wei Sun
Department of Mechanical Engineering and Mechanics, Drexel University, Philadelphia, PA, USA; Mechanical Engineering and Biomanufacturing Research Institute, Tsinghua University, Beijing, China; Shenzhen Biomanufacturing Engineering Laboratory, Shenzhen, Guangdong, China

Chengyang Wang
Department of Mechanical Engineering and Mechanics, Drexel University, Philadelphia, PA, USA

Paul S. Wiggenhauser
Department of Plastic and Hand Surgery, Klinikum Rechts der Isar, Technical University Munich, Germany

Yu Zhao
Mechanical Engineering and Biomanufacturing Research Institute, Tsinghua University, Beijing, China

Pinar Zorlutuna
Center for Biomedical Engineering, Department of Medicine, Brigham and Women's Hospital, Harvard Medical School, Cambridge, MA, USA; Harvard-MIT Division of Health Sciences and Technology, Massachusetts Institute of Technology Cambridge, MA, USA

In Vitro Biofabrication of Tissues and Organs

Koichi Nakayama

Graduate School of Science and Engineering, Saga University, Japan

CONTENTS

INTRODUCTION

After the sensational images of the mouse growing a human ear were broadcast around the world in the late 1990s, the in vitro fabrication of tissues and the regeneration of internal organs were no longer regarded as science fiction but as possible remedies for the millions suffering from chronic degenerative diseases.

Although some mistook it as a genetically engineered mouse expressing a human ear [1], these striking images nonetheless highlighted the medical promise of "tissue engineering" and ignited widespread interest from researchers in many fields, including cell and molecular biology, biomedical engineering, transplant medicine, and organic chemistry.

While there have already been successful clinical reports documenting the treatment of severe burn patients with culture-expanded skin cell sheets since the introduction of this tissue engineering technology in 1981 [2], fabrication of three-dimensional (3D) tissue constructs in vitro remains a challenge.

In the above-mentioned study, Cao et al. prepared a biodegradable polymer scaffold in the shape of a human ear and seeded its surface with bovine chondrocytes. This "tissue engineered ear" was then implanted under the skin of a nude mouse. As nutrients were provided by the in vivo environment, the implanted chondrocytes gradually started producing extracellular matrix (ECM) components such as collagen and glycoproteins. While a cell-free ear-shaped polymer could not have maintained its original shape in vivo due to the hydrolytic degradation of the polymer, the chondrocytes seeded onto the polymer maintained the original scaffold shape for 12 weeks after implantation. Indeed, the geometry was similar to and as complex as the original human ear.

After the study of the mouse with the human ear, many researchers attempted to create tissues or organs in vitro by constructing scaffolds composed of various biocompatible materials, such as animal-derived collagen [3], synthetic polymers [4], artificially synthesized bone substitutes (calcium-phosphate cement) [5], and autologous fibrin glue [6]. These scaffolds were seeded with a large array of somatic cells or stem cells to reconstruct target tissues such as skin [7], bladder [8], articular cartilage [9], liver [10], bone [11], vascular vessels [12], and even a finger [13].

The combination of a scaffold with cells and/or growth factors became the gold standard of tissue engineering [14]. Successful application of scaffold-based tissue engineering depends on three steps: (1) finding a source of precursor or stem cells from the patient, usually through biopsy or isolated from accessible stem cell-rich tissues, (2) seeding these cells in vitro onto scaffold material of the desired shape (with or without growth factors) that promotes cell proliferation, and (3) surgically implanting the scaffold into the target (injured) tissue of the patient.

This tissue engineering method overcomes a number of problems associated with allogeneic organ transplantation: the perpetual shortage of donors, the possibility of rejection, ethical issues such as organ trafficking [15], and the need for prolonged immunosuppression, which may lead to opportunistic infections and increased risk of cancer [16].

Many researchers tried to fabricate organs by combining cells, proteins/genes, and scaffolds. The various biomaterials used to fabricate scaffolds are classified into three types: (1) porous materials composed of biodegradable polymers, such as polylactic acid, polyglycolic acid, hyaluronic acid, and various co-polymers; (2) hydroxyapatite or calcium phosphate–based materials; and (3) soft materials like collagens, fibrin, and various hydrogels and their combinations.

In addition to providing a 3D structure for transplanted cells, scaffolds also dramatically enhance cell viability (e.g., a few exogenous cells were detected after the transplantation of single isolated cells into infarcted myocardium [17,18]). Anchorage-dependent cells cannot survive for long when detached from the surrounding ECM or culture surface. When there is loss of normal cell–cell and cell–ECM interactions, unanchored cells may undergo a specific form of programmed cell death called "anoikis" [19,20]. Thus, seeding anchorage-dependent cells onto scaffolds allows for efficient transplantation, especially if scaffolds are pretreated with growth factors. Indeed, some scaffold-based tissue engineered systems, such as bladder [21], articular cartilage [22], epidermis [23], and peripheral pulmonary arteries [24], have already been translated into the clinical stage.

1.1 Problems with scaffold-based tissue engineering

The ideal biodegradable scaffold polymer should be (1) nontoxic; (2) capable of maintaining mechanical integrity to allow tissue growth, differentiation, and integration; (3) capable of controlled degradation; and (4) nonimmunogenic; also, it should not cause infection or a prion-like disease. Although there are many clinical reports on the successful use of various biomaterials, there is still no "ideal" biomaterial for scaffold construction. Furthermore, concerns such as immunogenicity, long-term safety of scaffold degradation products, and the risk of infection or transmission of disease, either directly or concomitant with biofilm formation, remain to be resolved.

1.1.1 Immune reactions

A serious concern is that scaffolds may induce undesirable immune reactions [25], including inflammation, acute allergic responses, or late-phase responses. Scaffolds might even stimulate an autoimmune response, such as that produced by type II collagen in mice [26–28] used as models for rheumatoid arthritis. Immune responses may also be triggered by scaffold degradation byproducts. Metallosis is a specific form of inflammation induced by tiny metal particles that are shed from the metallic components of medical implants, such as debris from artificial joint prostheses [29]. Accumulation of scaffold degradation byproducts may elicit chronic diseases associated with inflammatory responses.

1.1.2 Degradation of scaffolds in vivo

Classic biodegradable polymers are defined as materials that are gradually digested by environmental bacteria through a process that is distinct from physiological degradation processes like digestion. Biodegradation can lead to toxicity in two ways: either a degradation product is directly toxic or it is metabolized to a toxic product (i.e., by liver enzymes). "Biodegradable" is distinct from "biocompatible." In most industrialized countries, only certified biomaterials that have passed multiple tests for severe toxicity and safety are permitted for use as medical implants.

Most synthesized biodegradable polymers are broken down by hydrolysis, resulting in the accumulation of acids that may alter the pH of the microenvironment or exert more direct toxicity. Some scaffolds are destroyed by macrophages, inducing an inflammatory reaction.

While bone substitute scaffolds may be replaced gradually by true bone through the activity of osteoclasts and osteoblasts, degradation of most other biomaterial scaffolds will leave a potential space that can impede repair. Biodegradable biomaterials are used extensively for cartilage repair, since articular cartilage (hyaline cartilage) has a low regenerative capacity and is usually replaced by weaker, rougher fibrous cartilage after injury [30]. When the scaffold is degraded and disappears, the space that once occupied it may no longer be filled with chondrocytes due to the cells' low proliferative capacity. These spaces might eventually form tiny cracks that trigger further deterioration of the smooth cartilage surface.

1.1.3 Risk of infection

There are two potential sources of infection from implanted scaffolds: pathogens transmitted directly from the scaffold or cells and infections emerging from the bacterial biofilm formed around the scaffolds after implantation.

1.1.3.1 Potential risk of disease transmission by scaffolds

Some scaffolds, such as collagen gels and amniotic membranes, are animal-derived. Recent outbreaks of severe infectious diseases like bovine spongiform encephalopathy and severe acute respiratory syndrome highlight the fact that animals harbor pathogens that may be lethal or cause severe infections in humans. Moreover, it is safe to assume that there are many undiscovered animal pathogens with the potential to cause human disease or death. Preclinical studies may minimize this risk, but there is no guarantee that these materials do not harbor unknown human pathogens.

1.1.3.2 Biofilms

Another source of infection from implanted scaffolds is the biofilm that forms on the scaffold surface [31]. Medical devices and implants, such as catheters and orthopedic or dental implants, are now ubiquitous in clinical practice. However, as the number of devices and implants continues to increase, the frequency of device-related infections will also increase [32,33]. Infections that are mostly caused by staphylococci, such as methicillin-resistant *Staphylococcus aureus*, usually do not respond to antibiotic therapy, necessitating removal of the implanted device.

In vivo microbial contamination of these devices differs from infection of natural tissues. Medical devices lack an immune system or bloodstream. Thus, once microorganisms invade through skin scratches, wounds, airways, or medical interventions and attach to the surface of the implanted material, they begin to form a bacterial biofilm [34]. The biofilm is composed of glycoproteins and polysaccharides secreted by microorganisms. Unlike circulating bacteria, biofilm-protected microorganisms are resistant to physical removal, host immunity, and antibiotics. Furthermore, since most antibiotics are unable to completely diffuse inside the biofilm, long-term antibiotic treatment may increase the risk of antibiotic resistance. In the United States, for example, catheter-related infections are a major cause of nosocomial morbidity and mortality. More than 300,000 U.S. patients are infected annually during presurgical or surgical procedures [35]. Moreover, as biofilms are slow to develop, infections due to biofilms may emerge several years after implantation. In artificial joint replacement surgery [36], this type of infection is a serious complication that can usually be cured only by removing the implant [32,37]. Infection by microorganisms is also widespread among contact lens users. One common cause of vision loss is contact lens—related microbial keratitis [38,39], and the risk of microbial keratitis increases during extended wear. This is why clinicians recommend frequent removal or replacement of contact lenses [38]. Furthermore, infection is the most common reason for breast implant removal [40,41]. These biofilm-related infections prolong hospitalization, increase medical costs, and sometimes result in mortality.

It is evident from the preceding discussion that scaffolds have several potential disadvantages. However, because there have been no clinical case reports documenting scaffold-related infection in regenerative medicine, many researchers have paid little attention to the possibility of infection from pathogens in the implant or biofilm.

Although most biomaterials used as scaffolds are biodegradable, degradation is usually very slow and may take several years. When infection occurs at the scaffold site, curing the infection may require surgical removal of the scaffold, disrupting tissue repair or causing further damage.

Various attempts have been made to develop infection-resistant biomaterials, such as silver ion—coated materials, ceramics that slowly release antibiotics [42], and antibacterial adhesion polymers [43], but it may take years before these materials are used in regenerative medicine, especially because these antibacterial factors may also harm the implanted cells. Thus, while scaffolds may hold great clinical potential, there remain significant safety concerns.

1.2 "Scaffold-free" tissue engineering

A precise definition of "scaffold-free" is still controversial [44]. Some investigators would insist that some of the techniques described below should not be called "scaffold-free" because the implanted construct may include residual biomaterials from the fabrication process. For the purpose of this section, a "scaffold-free" system is a "cell-only" construct that may or may not use other biomaterials during fabrication. Even if it does contain other biomaterials, these are not implanted along with the cells.

From a clinical perspective, the most important property of a scaffold is its behavior in the body upon implantation (degradation, biofilm formation) and physiological reactions induced by the parent material and degradation byproducts (immune responses, local or systemic infections).

1.2.1 Classification of present scaffold-free systems

Several scaffold-free systems have been reported, some of which are already used for clinical treatments. These systems can be divided into three categories according to the cellular material used for construction. One system uses single cell sheets, another uses isolated single cells, and the third uses spheroid cell aggregates as the essential building blocks for implantable 3D constructs (Figure 1.1).

1.2.1.1 Cell sheets

Cell sheet technology is one of the most advanced methods for the construction of implantable engineered tissue. Certain types of cells can be removed from a culture dish as a relatively stable confluent monolayer-sheet [45]. Cell sheet technology is already used clinically for the repair of skin [2], cornea [46], esophagus [47], heart muscle [48], and blood vessels [49], and it is a promising method for many other applications in tissue engineering and regenerative medicine.

The first successful clinical application of cell sheets was developed by Rheinwald and Green [45] to treat patients with severe burns. At that time, keratinocytes were difficult to culture for expansion. Rheinwald et al. seeded a suspension of disaggregated keratinocytes onto a feeder layer of irradiated mouse 3T3 cells. The feeder layer enhanced plating efficiency and stimulated keratinocyte proliferation. Proliferation and culture life span could be further increased by adding various supplements or growth factors to the culture medium. They were able to recover single continuous sheets of keratinocytes that could be grafted onto the sites of severe burns. Many patients with severe burns have survived due to this skin sheet technology [7]. Since then, grafting of these keratinocyte monolayers is perhaps the most successful example of tissue engineering therapy, and several products have been examined in clinical trials. A number of them have been approved by the FDA and are now on the market [50].

In January 2009, Japan Tissue Engineering Co., Ltd., a Japanese biotechnology company, began marketing autologous cultured epidermis (called JACE) as the first Japanese tissue engineering product covered by national health insurance. JACE uses Green's [45] cell sheet engineering system, and it is the only regenerative medicine product currently approved by the Japanese Ministry of Health, Labor and Welfare [51]. This approval is significant because the Japanese MHLW was considered to be the utmost conservative authority for the approval

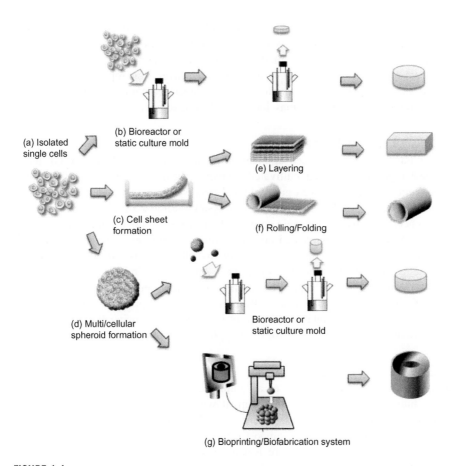

FIGURE 1.1

Methods for scaffold-free biofabrication. (a) Isolated single cell suspension. (b) Culture isolated cells in bioreactor or in static culture mold. (c) Cell sheet formation. (d) Multicellular cellular spheroid formation. (e) Layering cell sheets. (f) Rolling or folding cell sheets. (g) Computer-based bio-printing/biofabrication system.

of new drugs and medical devices and thus may indicate more timely approval and acceptance of similar products in Japan and elsewhere.

Okano et al. developed an alternative method for cell sheet engineering by first coating culture dishes with a temperature-responsive polymer, poly(N-isopropylacrylamide) [52]. This surface is relatively hydrophobic and similar to standard culture dishes at 37°C, but it becomes hydrophilic below 32°C. Various cell types can attach to the surface and proliferate at 37°C, while cooling below 32°C causes the cells to detach without the use of enzyme digestion reagent [53]. This is in contrast to Green's [45] cell sheet method, which always requires dispase for recovery of cell sheets from culture dishes. The method of Okano allows the production of many types of cell sheets that are too fragile or otherwise difficult to recover by other methods [53−55]. Furthermore, Okano's method does not require an earlier used exogenous feeder layer, thus representing a potentially safer method. (Earlier employed feeder layers containing mouse 3T3 cells produce mouse proteins that may induce allergic reactions.)

1.2.1.1.1 Corneal sheets

The clinically most advanced application of the system developed by Okano is corneal regeneration using cultivated human corneal sheet transplantation [46]. Kinoshita et al. also showed good clinical results with cultivated human corneal sheet transplantation [56]. However, their system is not scaffold-free by our definition because it used allogeneic amniotic human membrane as an autologous cell carrier. Nishida et al. harvested corneal epithelial stem cells from the limbus of patients with severe ocular trauma, such as alkali burns, or ocular diseases, including autoimmune disorders or Stevens-Johnson syndrome (erythema multiforme). After monolayer expansion in vitro, the corneal epithelial stem cells were formed into cell sheets using Okano's thermal responsive culture plates. Harvesting and transplantation of noninvasive cell sheets using this temperature-responsive culture system has also been applied for ocular surface regeneration.

1.2.1.1.2 Heart regeneration

Using Okano's method, Sawa et al. implanted a cultured skeletal muscle cell sheet into the damaged heart of a patient with degenerative cardiomyopathy, a disease characterized by progressive heart failure [48]. The patient was at end-stage heart failure and on life support using a mechanical left ventricle assisting system. The implanted cells were isolated from an approximately 10-g piece of skeletal muscle excised from the medial vastus muscle under general anesthesia. After monolayer expansion, 20 skeletal myoblast cell sheets were obtained and autologously implanted onto the patient's dilated heart through left lateral thoracotomy. Seven months after implantation, the patient was discharged from the hospital and no longer required artificial heart support.

1.2.1.1.3 Esophageal ulcer treatment

With the rapid progress of endoscopy, large esophageal cancers can be removed by a single procedure, such as endoscopic submucosal dissection (ESD). Endoscopic resection has become the standard treatment for esophageal lesions, replacing longer open surgery procedures. However, massive resection of esophageal cancer by ESD can be complicated by postoperative inflammation and stenosis (narrowing of the esophagus). Severe inflammation causes esophageal scarring, while stenosis may make swallowing difficult and painful. Although treatment with balloon dilation or temporary stents can enlarge the narrowed esophagus and partially and temporarily overcome this problem, an extended response generally requires repeated treatment that can lead to further inflammation and postoperative pain.

Postoperative inflammation and stenosis are caused mainly by massive ulceration of the esophageal surface. Following a successful preclinical trial in dogs, Ohki et al. performed clinical trials using cell sheets to treat large esophageal ulcers [57]. To this end, they developed a device that can directly transfer and attach cell sheets through endoscopy without suturing. The application of these epithelial cell layers enhanced wound healing and reduced host inflammatory responses without causing stenosis.

1.2.1.1.4 Blood vessels
Good clinical results were obtained when cell sheet–based scaffold-free blood vessels were used for the facilitation of hemodialysis (HD) treatment [12,58]. These vessels were fabricated by wrapping a dehydrated fibroblast sheet around a polytetrafluoroethylene (PTFE) tube cylinder and then overlaying a living smooth muscle cell sheet and an outer fibroblast sheet. After culturing this multi-layered "cell roll," the PTFE cylinder was removed and the lumenal surface was seeded with endothelial cells. A small clinical trial was conducted using this scaffold-free tube in ten patients with end-stage renal failure treated by HD through an arteriovenous fistula (shunt). Patients who require HD always face shunt complications such as infection and low blood flow due to clotting. L'Heureux et al. fabricated tissue engineered blood vessels with autologous cells from each patient and implanted the vessel as a replacement HD shunt. The implanted grafts were stable in vivo for 3 months and withstood repeated puncture for HD for up to 13 months, allowing uninterrupted HD [59].

1.2.1.1.5 Nerve grafts
Baltich et al. fabricated scaffold-free tubular constructs consisting of an external fibroblast layer and an internal core of interconnected neuronal cells derived from fetal rat spine. The conduction velocity of this engineered "nerve" was comparable to that of the sciatic nerve of a 4-week-old rat and approximately 50 percent of that observed in a 12-week-old (adult) rat [60]. These results suggest that the scaffold-free nerve grafts may be useful for peripheral nerve repair.

1.2.1.1.6 Liver regeneration
Fabrication of liver tissue in vitro has attracted considerable interest given the innate regenerative capacity of the liver and prevalence of liver diseases. Ohashi et al. [10] fabricated hepatocyte cell sheets by culturing hepatocytes on temperature-responsive poly(N-isopropylacrylamide)-coated culture dishes. Sheets of hepatic tissue transplanted ectopically into the subcutaneous space were pretreated with growth factor FGF-2 to promote neovascularization. These sheets efficiently integrated with the surrounding tissue and persisted for longer than 200 days. The engineered and implanted hepatic tissues also showed several characteristics of liver-specific functionality. Furthermore, layered hepatic tissue

sheets reorganized into a 3D structure with the histological appearance of liver tissue [10].

1.2.1.1.7 Implantation of pancreatic islets

The observed functional differentiation of ectopically implanted hepatocytes led Ohashi and Okano to perform a similar experiment using cell sheets composed of rat pancreatic islet cells [61]. In vitro, these pancreatic islet cell sheets retained the functional activity of native islet cells, including the production of insulin and glucagon, and glucose-dependent insulin secretion. Moreover, when transplanted into the subcutaneous space of rats, pancreatic islet cell sheets produced and secreted insulin, suggesting a new therapeutic approach for the treatment of diabetes mellitus and other diseases involving dysfunctional islet cells and possible elimination of the need for daily insulin injections [62,63].

The Okano group is now aggressively expanding potential applications by developing sheets for regeneration of bone [64], articular cartilage surface [65], periodontal ligament [66], lung [67], thyroid [68], and bladder [69].

1.2.1.1.8 Expansion of cell sheets into 3D structures

Various approaches have been used to fabricate larger 3D tissues and organs from cell sheets. One example is the "Origami" approach, where, like L'Heureux et al. [12], sheets are formed around a temporary 3D scaffold (like surgical tubing). Another standard approach is the layering of multiple cell sheets. Shimizu et al. layered beating cardiomyocyte sheets to fabricate scaffold-free 3D constructs and found that these layered cardiomyocytes exhibited synchronized beating. However, the maximum thickness was limited to less than 80 μm (three-cell layer), possibly due to starvation and hypoxia of inner layers that have poor access to the culture media and atmosphere. Moreover, cardiomyocytes are tightly interconnected by gap junctions, and the outer layer may prevent gas and nutrient exchange to the inner layers. Indeed, four-layered cardiomyocyte constructs showed necrosis in the inner layers [70]. To overcome this limited maximum thickness, they implanted 10 three-layered cardiomyocyte sheets into nude rat hearts at 1- or 2-day intervals, finally obtaining a 1-mm-thick neomyocardium fused onto the native heart and integrated with a well-organized microvascular network. Although it is obviously impossible to perform multiple thoracotomies on human patients, this demonstration revealed the importance of neovascularization for the gradual construction of larger cell constructs with or without the use of scaffolds [71,72].

1.2.2 In vitro self-produced ECM-rich scaffold-free constructs

Certain cell types possess the capacity to synthesize and release components of ECM in vitro under appropriate culture conditions. Fibroblasts and chondrocytes, for example, produce collagens and proteoglycans in vitro. This ECM production capacity is accelerated under confluence or 3D culture conditions, a phenomenon that may be inspired to develop scaffold-free systems for fabrication of 3D constructs from isolated chondrocytes [73,74]. Normal anchorage-dependent cells proliferate at low density in monolayer culture. When the cell density reaches a certain threshold, proliferation is suppressed by contact inhibition. Under contact inhibition, the cell cycle stops and chondrocytes start to produce ECM proteins. Using this in vitro−produced ECM, many groups have developed methods to fabricate 3D scaffold-free constructs in vitro. Most of these approaches are used for fabrication of cartilage-like constructs [75,76]. In this technique, a large number of isolated chondrocytes is loaded into a specific culture mold and fed reagents that enhance matrix production. Since chondrocytes exist under relatively low oxygen partial pressure and without blood supply in vivo (in joints), they are relatively harder than normal cells and can be cultured under high-density static culture conditions. Although these approaches require relatively longer culture, the resulting constructs are similar to native cartilage in terms of histology and biomechanical properties [77−79]. However, these "in vitro self-produced ECM-rich scaffold-free" constructs have limitations as well. It is difficult to expand cultures in 3D without hypoxia or nutrient starvation of inner core cells. Thus, most of these scaffold-free cartilage-like constructs are thinner than normal human articular cartilage in adults.

1.2.3 The rotating wall vessel bioreactor system

Another approach for fabricating scaffold-free constructs from isolated cells is by using a rotating wall vessel bioreactor system. This culture system utilizes a circular vessel with a gas-exchange membrane and rotates around a horizontal axis to provide culture media flow in a simulated microgravity environment. The rotating wall reactor (RWR) was developed by NASA to produce cartilage-like tissue in space [80]. This reactor has also been used on earth to fabricate various other cell constructs [81−83]. Okamura et al. loaded isolated hepatocytes into an RWR and obtained a "liver-like" construct with bile duct− and vessel-like structures formed within the tissue. Histological analysis showed that the bile duct structures secreted mucin and formed complex tubular branches in the peripheral region. Distal to these bile duct structures, they observed mature hepatocytes capable of producing albumin and storing glycogen [84]. To our knowledge, there are still no clinical reports using engineered tissue fabricated by this method.

1.3 Aggregation/spheroid-based approaches

The capacity of dissociated cells to reaggregate through cell−cell attachment has been known for over 100 years [85]. This phenomenon is preserved in almost all living organisms irrespective of their complexity [86,87]. These aggregates are usually called multicellular spheroids (MCSs) and are powerful research tools in modern developmental biology, stem cell biology, tumor biology, toxicology, and pharmacology.

1.3.1 Preparation of multicellular spheroids

Although several methods for MCS preparation have been introduced, they all rely on a simple common principle: dissociated cells are incubated in a nonadhesive environment to allow individual cells to attach to one another. This cell–cell attachment is a survival mechanism that allows cells to avoid anoikis, possibly by activating signals mediated by surface receptors and ligands that suppress the anoikis cascade.

These MCSs can be prepared in regular nonadhesive culture dishes, silicon-coated dishes [88], containers coated with nonadhesive enhanced polymers (such as PDMS [89]) or poly-HEMA [90], agarose gels [91], alginate beads [92], spinner flasks [78], or hanging drop cultures [93]. After reaggregation of dissociated cells, each MCS has the capacity to fuse with other MCSs. Many groups have developed alternative approaches for scaffold-free tissue engineering using this propensity for MCS fusion. In fact, methods using spheroids as building blocks for fabrication of scaffold-free cell constructs may be a better approach because many cells in spheroids show greater similarity to cells in the native state than do cells in monolayer culture [94].

MCS fusion usually requires 24 to 72 hours, depending on the cell type and culture conditions. During fusion, these MCSs must be kept in culture media under controlled conditions because even a slight tilting of the culture dish may deform the desired shape of the stacked MCS blocks.

1.3.2 Molding MCSs

Most of these MCS-based approaches also use specific molding chambers to produce constructs of the desired shapes [91,95,96]. However, it remains difficult to fabricate larger tissues similar to the geometry of native tissues or organs due to limited gas and nutrient exchange within the core of the spheroids, particularly for hepatocytes, cardiomyocytes, and other cells with high nutritional or metabolic demands.

1.3.3 Bio-printing

Possibly inspired by common inkjet printers, cell printing systems, called bio-printing systems, have been developed [71,97]. Mironov, Forgacs, and colleagues first established a spheroid-based bio-printing system [98,99] that used MCSs as "bio-ink" and hydrogels as "bio-paper." Their printer lays down MCSs onto pre-designed spots on the hydrogel to allow adjacent MCSs to fuse until the desired shape is attained. Using this system, they fabricated a beating cardiomyocyte plate. Their latest system can print multicellular rods onto agarose (bio-paper) using a dual-nozzle system for real-time molding of vascular and peripheral nerve constructs [100]. This bio-printing approach can fabricate more complex three-dimensional designs with microchannels that may allow for better in vitro perfusion and provide conduits for neovasculization in vivo.

1.3.4 Alternative approach for MCS assembly technique for biofabrication

We developed another approach for a scaffold-free MCS assembling system called a "needle-array" system (Figure 1.2) that is slightly different from bioprinting systems. Instead of using hydrogel as the "bio-paper," we used medical-grade stainless needles as temporal fixators to skewer MCSs until they fused with one another. This concept was inspired from surgical treatments for bone fracture in orthopedic surgery, called "external fixation" (Figure 1.3). For treatment of bone fracture, orthopedic surgeons reposition fractured bone pieces to their original positions with or without surgery. After repositioning, surgeons immobilize bone pieces by using casts or splints without surgery or by using metal plates, screws, or pins under surgical procedure. Inspired from fracture treatment, especially by the external fixation technique, we developed the needle-array

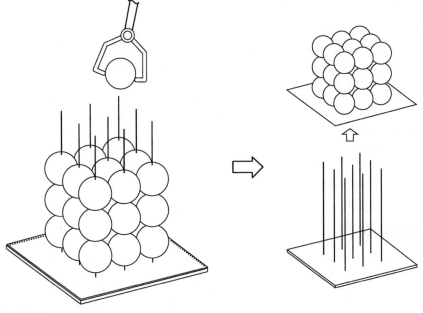

FIGURE 1.2

The needle-array system. (Left) Skewering a spheroid into the needle-array with a robotic system. (Right) Removing fused spheroids to obtain a scaffold-free construct.

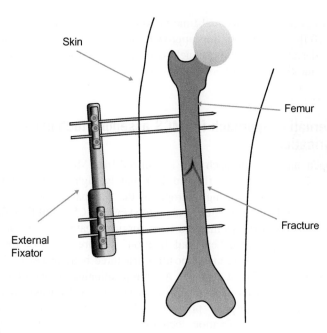

FIGURE 1.3

External fixation for bone fracture treatment. Pins are inserted through the skin into the bone.

system. We also developed a robotic system that skewers MCSs into needle-array according to a three-dimensional design template (Figure 1.4).

By applying these systems, we can fabricate complex three-dimensional scaffold-free cell constructs by using various types of cells such as chondrocyte, hepatocyte, cardiomyocyte, vascular smooth muscle cell, and so on. Since we utilize medical grade needles, it is easy to remove the temporary supports without contamination with exogenous materials. In addition, efficient gas and nutrition exchange could be expected, unlike in other scaffold-free MCS-based systems.

CONCLUSION

In this section, we reviewed the pros and cons of tissue engineering using scaffolds for regeneration of damaged tissue and discussed current developments in scaffold-free systems, some of which have already found clinical applications. We possess a large array of tissue engineering/regenerative medicine systems that must now be refined for clinical applications. Some of these systems show remarkable results in vitro but are challenging to translate into bedside.

FIGURE 1.4

Needle-array-based scaffold-free biofabrication system. (Left) DNA-like double helix design template. (Middle) 3D biofabrication system in a bio-clean bench. (Right) Scaffold-free endothelial cell—based construct based on the design template in the left pane. Scale bar = 1 mm.

The industrial giant Dow Corning, the largest supplier of silicone breast implants, filed for Chapter 11 bankruptcy in 1995 in the face of over 20,000 lawsuits claiming that its implants caused systemic health problems, despite the fact that there was no direct evidence linking the implants to these health problems [101]. The lesson from this case is that despite promising results in vitro and in preclinical models (that are usually only monitored for a few months or years), the safety of scaffolds and implants must be confirmed in humans over the long term. Scaffold-free systems are an alternative for tissue engineering and repair of damaged tissues that may circumvent at least some of these potential risks.

Tissue engineering and regenerative medicine are under constant development, so it is too early to determine which is better: scaffold or scaffold-free engineered tissues. Whichever approach is used, to fabricate human-scale tissues or organs in vitro, new methods must allow for neovascularization to overcome the diffusion limits of oxygen and nutrients within tissues [71,72]. Endothelial cells have the capacity to form tubes and networks in vitro under appropriate culture conditions [102], so there is hope that the problems of neovascularization can be solved without the need for fabricating complex vascular networks in vitro.

In light of the latest developments on decellularized organs [103,104], it can be surmised that it will soon be possible to fabricate whole tissues and organs in vitro using the appropriate combination of cells, culture conditions, and bioreactors without the use of artificial scaffolds. To achieve the translation of these emerging tissue engineering technologies from benchside to bedside, with or without a scaffold, the safety, efficacy, and cost-effectiveness of the approach will have to be evaluated at each phase of clinical development.

References

[1] Cao Y, Vacanti JP, Paige KT, Upton J, Vacanti CA. Transplantation of chondrocytes utilizing a polymer-cell construct to produce tissue-engineered cartilage in the shape of a human ear. Plast Reconstr Surg 1997;100(2):297−302.

[2] O'Connor N, Mulliken J, Banks-Schlegel S, Kehinde O, Green H. Grafting of burns with cultured epithelium prepared from autologous epidermal cells. Lancet 1981;317 (8211):75−8.

[3] Wakitani S, Kimura T, Hirooka A, Ochi T, Yoneda M, Yasui N, et al. Repair of rabbit articular surfaces with allograft chondrocytes embedded in collagen gel. J Bone Joint Surg Br 1989;71(1):74−80.

[4] Shin'oka T, Imai Y, Ikada Y. Transplantation of a tissue-engineered pulmonary artery. N Engl J Med 2001;344(7):532−3.

[5] Oreffo RO, Driessens FC, Planell JA, Triffitt JT. Growth and differentiation of human bone marrow osteoprogenitors on novel calcium phosphate cements. Biomaterials 1998;19(20):1845−54.

[6] Fussenegger M, Meinhart J, Höbling W, Kullich W, Funk S, Bernatzky G. Stabilized autologous fibrin-chondrocyte constructs for cartilage repair in vivo. Ann Plast Surg 2003;51(5):493−8.

[7] Gallico GG, O'Connor NE, Compton CC, Kehinde O, Green H. Permanent coverage of large burn wounds with autologous cultured human epithelium. N Engl J Med 1984;311(7):448−51.

[8] Kropp BP, Cheng EY. Bioengineering organs using small intestinal submucosa scaffolds: in vivo tissue-engineering technology. J Endourol 2000;14(1):59−62.

[9] Brittberg M, Nilsson A, Lindahl A, Ohlsson C, Peterson L. Rabbit articular cartilage defects treated with autologous cultured chondrocytes. Clin Orthop Relat Res 1996;326:270−83.

[10] Ohashi K, Yokoyama T, Yamato M, Kuge H, Kanehiro H, Tsutsumi M, et al. Engineering functional two-and three-dimensional liver systems in vivo using hepatic tissue sheets. Nat Med 2007;13(7):880−5.

[11] Niemeyer P, Krause U, Fellenberg J, Kasten P, Seckinger A, Ho AD, et al. Evaluation of mineralized collagen and alpha-tricalcium phosphate as scaffolds for tissue engineering of bone using human mesenchymal stem cells. Cells Tissues Organs 2004;177(2):68−78.

[12] L'Heureux N, Pâquet S, Labbé R, Germain L, Auger FA. A completely biological tissue-engineered human blood vessel. FASEB J 1998;12(1):47−56.

[13] Isogai N, Landis W, Kim TH, Gerstenfeld LC, Upton J, Vacanti JP. Formation of phalanges and small joints by tissue-engineering. J Bone Joint Surg Am 1999;81 (3):306−16.

[14] Langer R, Vacanti JP. Tissue engineering. Science 1993;260(5110):920−6.

[15] The declaration of Istanbul on organ trafficking and transplant tourism. Clin J Am Soc Nephrol 2008;3(5):1227−31.

[16] Pollard JD, Hanasono MM, Mikulec AA, Le QT, Terris DJ. Head and neck cancer in cardiothoracic transplant recipients. Laryngoscope 2000;110(8):1257−61.

[17] Zhang M, Methot D, Poppa V, Fujio Y, Walsh K, Murry CE. Cardiomyocyte grafting for cardiac repair: graft cell death and anti-death strategies. J Mol Cell Cardiol 2001;33(5):907−21.

[18] Hofmann M, Wollert KC, Meyer GP, Menke A, Arseniev L, Hertenstein B, et al. Monitoring of bone marrow cell homing into the infarcted human myocardium. Circulation 2005;111(17):2198−202.

[19] Frisch SM, Ruoslahti E. Integrins and anoikis. Curr Opin Cell Biol 1997;9(5):701−6.

[20] Gilmore AP. Anoikis. Cell Death Differ 2005;12(Suppl 2):1473−7.

[21] Atala A. Experimental and clinical experience with tissue engineering techniques for urethral reconstruction. Urol Clin North Am 2002;29(2):485−92, ix.

[22] Behrens P, Bitter T, Kurz B, Russlies M. Matrix-associated autologous chondrocyte transplantation/implantation (MACT/MACI)–5-year follow-up. Knee 2006;13(3): 194−202.

[23] Galassi G, Brun P, Radice M, Cortivo R, Zanon GF, Genovese P, et al. In vitro reconstructed dermis implanted in human wounds: degradation studies of the HA-based supporting scaffold. Biomaterials 2000;21(21):2183−91.

[24] Hibino N, Imai Y, Shin-oka T, Aoki M, Watanabe M, Kosaka Y, et al. First successful clinical application of tissue engineered blood vessel Kyobu geka. Jpn J Thorac Surg 2002;55(5):368−73.

[25] Schakenraad JM, Dijkstra PJ. Biocompatibility of poly (DL-lactic acid/glycine) copolymers. Clin Mater 1991;7(3):253−69.

[26] Billingham ME. Models of arthritis and the search for anti-arthritic drugs. Pharmacol Ther 1983;21(3):389−428.

[27] Wooley PH. Animal models of rheumatoid arthritis. Curr Opin Rheumatol 1991;3 (3):407−20.

[28] Matsuo A, Shuto T, Hirata G, Satoh H, Matsumoto Y, Zhao H, et al. Antiinflammatory and chondroprotective effects of the aminobisphosphonate incadronate (YM175) in adjuvant induced arthritis. J Rheumatol 2003;30(6):1280−90.

[29] Huo MH, Romness DW, Huo SM. Metallosis mimicking infection in a cemented total knee replacement. Orthopedics 1997;20(5):466−70.

[30] Patrascu JM, Freymann U, Kaps C, Poenaru DV. Repair of a post-traumatic cartilage defect with a cell-free polymer-based cartilage implant: a follow-up at two years by MRI and histological review. J Bone Joint Surg Br 2010;92(8):1160−3.

[31] Khardori N, Yassien M. Biofilms in device-related infections. J Ind Microbiol 1995;15(3):141−7.

[32] Van de Belt H, Neut D, Schenk W, Van Horn JR, Van der Mei HC, Busscher HJ. Infection of orthopedic implants and the use of antibiotic-loaded bone cements. A review. Acta Orthop Scand 2001;72(6):557−71.

[33] Rimondini L, Fini M, Giardino R. The microbial infection of biomaterials: a challenge for clinicians and researchers. A short review. J Appl Biomater Biomech 2005;3(1):1−10.

[34] Costerton JW, Stewart PS, Greenberg EP. Bacterial biofilms: a common cause of persistent infections. Science 1999;284(5418):1318−22.

[35] Gristina AG, Giridhar G, Gabriel BL, Naylor PT, Myrvik QN. Cell biology and molecular mechanisms in artificial device infections. Int J Artif Organs 1993;16(11):755−63.

[36] Fukagawa S, Matsuda S, Miura H, Okazaki K, Tashiro Y, Iwamoto Y. High-dose antibiotic infusion for infected knee prosthesis without implant removal. J Orthop Sci 2010;15(4):470−6.

[37] Garvin KL, Konigsberg BS. Infection following total knee arthroplasty: prevention and management. J Bone Joint Surg Am 2011;93(12):1167−75.

[38] Holden BA, Sweeney DF, Sankaridurg PR, Carnt N, Edwards K, Stretton S, et al. Microbial keratitis and vision loss with contact lenses. Eye Contact Lens 2003;29 (1 Suppl):[S131−4; discussion S143−4, S192−4].

[39] Stapleton F, Keay L, Edwards K, Naduvilath T, Dart JKG, Brian G, et al. The incidence of contact lens-related microbial keratitis in Australia. Ophthalmology 2008;115(10):1655−62.

[40] Virden CP, Dobke MK, Stein P, Parsons CL, Frank DH. Subclinical infection of the silicone breast implant surface as a possible cause of capsular contracture. Aesthetic Plast Surg 1992;16(2):173−9.

[41] Pittet B, Montandon D, Pittet D. Infection in breast implants. Lancet Infect Dis 2005;5(2):94−106.

[42] Shinto Y, Uchida A, Korkusuz F, Araki N, Ono K. Calcium hydroxyapatite ceramic used as a delivery system for antibiotics. J Bone Joint Surg Br 1992;74(4):600−4.

[43] Fu J, Ji J, Yuan W, Shen J. Construction of anti-adhesive and antibacterial multilayer films via layer-by-layer assembly of heparin and chitosan. Biomaterials 2005;26 (33):6684−92.

[44] Guillemot F, Mironov V, Nakamura M. Bioprinting is coming of age: report from the international conference on bioprinting and biofabrication in bordeaux (3B'09). Biofabrication 2010;2(1):010201.

[45] Rheinwald JG, Green H. Serial cultivation of strains of human epidermal keratinocytes: the formation of keratinizing colonies from single cells. Cell 1975;6(3):331−43.

[46] Nishida K, Yamato M, Hayashida Y, Watanabe K, Yamamoto K, Adachi E, et al. Corneal reconstruction with tissue-engineered cell sheets composed of autologous oral mucosal epithelium. N Engl J Med 2004;351(12):1187−96.

[47] Ohki T, Yamato M, Ota M, Murakami D, Takagi R, Kondo M, et al. Endoscopic transplantation of human oral mucosal epithelial cell sheets-world's first case of regenerative medicine applied to endoscopic treatment. Gastrointest Endosc 2009;69 (5):AB253−4.

[48] Sawa Y, Miyagawa S, Sakaguchi T, Fujita T, Matsuyama A, Saito A, et al. Tissue engineered myoblast sheets improved cardiac function sufficiently to discontinue LVAS in a patient with DCM: report of a case. Surg Today 2012;42(2):181−4.

[49] L'Heureux N, McAllister TN, De la Fuente LM. Tissue-engineered blood vessel for adult arterial revascularization. N Engl J Med 2007;357(14):1451−3.

[50] Phillips TJ. New skin for old: developments in biological skin substitutes. Arch Dermatol 1998;134(3):344−9.

[51] Japanese Ministry of Health, Labor and Welfare (J-TEC) (2009): J-TEC Top message. Retrieved am from <http://www.jpte.co.jp/english/ir/top_message.html>.

[52] Yamada N, Okano T, Sakai H, Karikusa F, Sawasaki Y, Sakurai Y. Thermoresponsive polymeric surfaces; control of attachment and detachment of cultured cells. Makromol Chem Rapid Comm 1990;11(11):571−6.

[53] Shimizu T, Yamato M, Kikuchi A, Okano T. Two-dimensional manipulation of cardiac myocyte sheets utilizing temperature-responsive culture dishes augments the pulsatile amplitude. Tissue Eng 2001;7(2):141−51.

[54] Nakajima K, Honda S, Nakamura Y, López-Redondo F, Kohsaka S, Yamato M, et al. Intact microglia are cultured and non-invasively harvested without pathological activation using a novel cultured cell recovery method. Biomaterials 2001;22 (11):1213−23.

[55] Okano T, Yamada N, Sakai H, Sakurai Y. A novel recovery system for cultured cells using plasma-treated polystyrene dishes grafted with poly(*N*-isopropylacrylamide). J Biomed Mater Res 1993;27(10):1243−51.

[56] Ishino Y, Sano Y, Nakamura T, Connon CJ, Rigby H, Fullwood NJ, et al. Amniotic membrane as a carrier for cultivated human corneal endothelial cell transplantation. Invest Ophthalmol Vis Sci 2004;45(3):800−6.

[57] Ohki T, Yamato M, Murakami D, Takagi R, Yang J, Namiki H, et al. Treatment of oesophageal ulcerations using endoscopic transplantation of tissue-engineered autologous oral mucosal epithelial cell sheets in a canine model. Gut 2006;55 (12):1704−10.

[58] L'Heureux N, Dusserre N, Konig G, Victor B, Keire P, Wight TN, et al. Human tissue-engineered blood vessels for adult arterial revascularization. Nat Med 2006;12 (3):361−5.

[59] McAllister TN, Maruszewski M, Garrido SA, Wystrychowski W, Dusserre N, Marini A, et al. Effectiveness of haemodialysis access with an autologous tissue-engineered vascular graft: a multicentre cohort study. Lancet 2009;373(9673):1440−6.

[60] Baltich J, Hatch-Vallier L, Adams AM, Arruda EM, Larkin LM. Development of a scaffoldless three-dimensional engineered nerve using a nerve-fibroblast co-culture. In Vitro Cell Dev Biol Anim 2010;46(5):438−44.

[61] Shimizu H, Ohashi K, Utoh R, Ise K, Gotoh M, Yamato M, et al. Bioengineering of a functional sheet of islet cells for the treatment of diabetes mellitus. Biomaterials 2009;30(30):5943−9.

[62] Ohashi K, Mukobata S, Utoh R, Yamashita S, Masuda T, Sakai H, et al. Production of islet cell sheets using cryopreserved islet cells. Transplant Proc 2011;43 (9):3188−91.

[63] Saito T, Ohashi K, Utoh R, Shimizu H, Ise K, Suzuki H, et al. Reversal of diabetes by the creation of neo-islet tissues into a subcutaneous site using islet cell sheets. Transplantation 2011;92(11):1231−6.

[64] Uchiyama H, Yamato M, Sasaki R, Sekine H, Yang J, Ogiuchi H, et al. In vivo 3D analysis with micro-computed tomography of rat calvaria bone regeneration using periosteal cell sheets fabricated on temperature-responsive culture dishes. J Tissue Eng Regen Med 2011;5(6):483−90.

[65] Kaneshiro N, Sato M, Ishihara M, Mitani G, Sakai H, Mochida J. Bioengineered chondrocyte sheets may be potentially useful for the treatment of partial thickness defects of articular cartilage. Biochem Biophys Res Commun 2006;349(2):723−31.

[66] Tsumanuma Y, Iwata T, Washio K, Yoshida T, Yamada A, Takagi R, et al. Comparison of different tissue-derived stem cell sheets for periodontal regeneration in a canine 1-wall defect model. Biomaterials 2011;32(25):5819−25.

[67] Kanzaki M, Yamato M, Yang J, Sekine H, Kohno C, Takagi R, et al. Dynamic sealing of lung air leaks by the transplantation of tissue engineered cell sheets. Biomaterials 2007;28(29):4294−302.

[68] Arauchi A, Shimizu T, Yamato M, Obara T, Okano T. Tissue-engineered thyroid cell sheet rescued hypothyroidism in rat models after receiving total thyroidectomy comparing with nontransplantation models. Tissue Eng Part A 2009;15(12):3943−9.

[69] Watanabe E, Yamato M, Shiroyanagi Y, Tanabe K, Okano T. Bladder augmentation using tissue-engineered autologous oral mucosal epithelial cell sheets grafted on demucosalized gastric flaps. Transplantation 2011;91(7):700−6.

[70] Shimizu T. Fabrication of pulsatile cardiac tissue grafts using a novel 3-Dimensional cell sheet manipulation technique and temperature-responsive cell culture surfaces. Circ Res 2002;90(3):40e−8e.

[71] Mironov V, Boland T, Trusk T, Forgacs G, Markwald RR. Organ printing: computer-aided jet-based 3D tissue engineering. Trends Biotechnol 2003;21(4):157−61.

[72] Sekiya S, Shimizu T, Yamato M, Okano T. "Deep-media culture condition" promoted lumen formation of endothelial cells within engineered three-dimensional tissues in vitro. J Artif Organs 2011;14(1):43−51.

[73] Park K, Huang J, Azar F, Jin RL, Min B-H, Han DK, et al. Scaffold-free, engineered porcine cartilage construct for cartilage defect repair−in vitro and in vivo study. Artif Organs 2006;30(8):586−96.

[74] Jin RL, Park SR, Choi BH, Min B-H. Scaffold-free cartilage fabrication system using passaged porcine chondrocytes and basic fibroblast growth factor. Tissue Eng Part A 2009;15(8):1887−95.

[75] Grogan S. A static, closed and scaffold-free bioreactor system that permits chondrogenesis in vitro. Osteoarthritis Cartilage 2003;11(6):403−11.

[76] Ando W, Tateishi K, Hart DA, Katakai D, Tanaka Y, Nakata K, et al. Cartilage repair using an in vitro generated scaffold-free tissue-engineered construct derived from porcine synovial mesenchymal stem cells. Biomaterials 2007;28(36):5462−70.

[77] Mainil-Varlet P, Rieser F, Grogan S, Mueller W, Saager C, Jakob RP. Articular cartilage repair using a tissue-engineered cartilage-like implant: an animal study. Osteoarthritis Cartilage 2001;9:S6−15.

[78] Nagai T, Furukawa KS, Sato M, Ushida T, Mochida J. Characteristics of a scaffold-free articular chondrocyte plate grown in rotational culture. Tissue Eng Part A 2008;14(7):1183−93.

[79] Miyazaki T, Miyauchi S, Matsuzaka S. Formation of proteoglycan and collagen-rich scaffold-free stiff cartilaginous tissue using two-step culture methods with combinations of growth factors. Tissue 2010;16(5).

[80] Freed LE, Langer R, Martin I, Pellis NR, Vunjak-Novakovic G. Tissue engineering of cartilage in space. Proc Natl Acad Sci USA 1997;94(25):13885−90.

[81] Barrila J, Radtke AL, Crabbé A, Sarker SF, Herbst-Kralovetz MM, Ott CM, et al. Organotypic 3D cell culture models: using the rotating wall vessel to study host-pathogen interactions. Nat Rev Microbiol 2010;8(11):791−801.

[82] Botta GP, Manley P, Miller S, Lelkes PI. Real-time assessment of three-dimensional cell aggregation in rotating wall vessel bioreactors in vitro. Nat Protoc 2006;1(4):2116−27.

[83] Sakai S, Mishima H, Ishii T, Akaogi H, Yoshioka T, Ohyabu Y, et al. Rotating three-dimensional dynamic culture of adult human bone marrow-derived cells for tissue engineering of hyaline cartilage. J Orthop Res 2009;27(4):517−21.

[84] Okamura A, Zheng Y-W, Hirochika R, Tanaka J, Taniguchi H. In-vitro reconstitution of hepatic tissue architectures with neonatal mouse liver cells using three-dimensional culture. J Nanosci Nanotechnol 2007;7(3):721−5.

[85] Wilson H. On some phenomena of coalescence and regeneration in sponges. J Exp Zool 1907;5(2):245−58.

[86] Townes PL, Holtfreter J. Directed movements and selective adhesion of embryonic amphibian cells. J Exp Zool 1955;128(1):53−120.

[87] Steinberg MS. Mechanism of tissue reconstruction by dissociated cells, II: time-course of events inine scene- Utilization of Nitrogen Compounds by Unicellular Algae 1954;137(X 127).

[88] Sakai Y, Naruse K, Nagashima I, Muto T, Suzuki M. Large-scale preparation and function of porcine hepatocyte spheroids. Int J Artif Organs 1996;19(5):294–301.

[89] Nishikawa M, Yamamoto T, Kojima N, Kikuo K, Fujii T, Sakai Y. Stable immobilization of rat hepatocytes as hemispheroids onto collagen-conjugated poly-dimethylsiloxane (PDMS) surfaces: importance of direct oxygenation through PDMS for both formation and function. Biotechnol Bioeng 2008;99(6):1472–81.

[90] Seidel JM, Malmonge SM. Synthesis of polyHEMA hydrogels for using as biomaterials. Bulk and solution radical-initiated polymerization techniques. Mater Res 2000;3(3):79–83.

[91] Dean DM, Napolitano AP, Youssef J, Morgan JR. Rods, tori, and honeycombs: the directed self-assembly of microtissues with prescribed microscale geometries. FASEB J 2007;21(14):4005–12.

[92] Masuda K, Sah RL, Hejna MJ, Thonar EJ-MA. A novel two-step method for the formation of tissue-engineered cartilage by mature bovine chondrocytes: the alginate-recovered-chondrocyte (ARC) method. J Orthop Res 2003;21(1):139–48.

[93] Kelm JM, Timmins NE, Brown CJ, Fussenegger M, Nielsen LK. Method for generation of homogeneous multicellular tumor spheroids applicable to a wide variety of cell types. Biotechnol Bioeng 2003;83(2):173–80.

[94] Chang TT, Hughes-fulford M. Monolayer and spheroid culture of human liver hepatocellular carcinoma cell line cells demonstrate and functional phenotypes. Dev Cell 2009;15(3).

[95] Kelm JM, Djonov V, Ittner LM, Fluri D, Born W, Hoerstrup SP, et al. Design of custom-shaped vascularized tissues using microtissue spheroids as minimal building units. Tissue Eng 2006;12(8):2151–60.

[96] Rago AP, Dean DM, Morgan JR. Controlling cell position in complex heterotypic 3D microtissues by tissue fusion. Biotechnol Bioeng 2009;102(4):1231–41.

[97] Boland T, Mironov V, Gutowska A, Roth EA, Markwald RR. Cell and organ printing 2: fusion of cell aggregates in three-dimensional gels. Anat Rec A Discov Mol Cell Evol Biol 2003;272(2):497–502.

[98] Jakab K, Neagu A, Mironov V, Forgacs G. Organ printing: fiction or science. Biorheology 2004;41(3-4):371–5.

[99] Jakab K, Neagu A, Mironov V, Markwald RR, Forgacs G. Engineering biological structures of prescribed shape using self-assembling multicellular systems. Proc Natl Acad Sci U S A 2004;101(9):2864–9.

[100] Norotte C, Marga FS, Niklason LE, Forgacs G. Scaffold-free vascular tissue engineering using bioprinting. Biomaterials 2009;30(30):5910–7.

[101] Renwick SB. Silicone breast implants: implications for society and surgeons. Med J Aust 1996;165(6):338–41.

[102] Arnaoutova I, George J, Kleinman HK, Benton G. The endothelial cell tube formation assay on basement membrane turns 20: state of the science and the art. Angiogenesis 2009;12(3):267–74.

[103] Ott HC, Matthiesen TS, Goh S-K, Black LD, Kren SM, Netoff TI, et al. Perfusion-decellularized matrix: using nature's platform to engineer a bioartificial heart. Nat Med 2008;14(2):213–21.

[104] Uygun BE, Soto-Gutierrez A, Yagi H, Izamis M-L, Guzzardi MA, Shulman C, et al. Organ reengineering through development of a transplantable recellularized liver graft using decellularized liver matrix. Nat Med 2010;16(7):814–20.

Biomaterials for Biofabrication of 3D Tissue Scaffolds

2

Joydip Kundu,[1] Falguni Pati,[1] Young Hun Jeong,[2] Dong-Woo Cho[3]

[1]Department of Mechanical Engineering, POSTECH, South Korea; [2]Department of Mechanical Engineering, Korea Polytechnic University, South Korea; [3]Department of Mechanical Engineering, POSTECH, South Korea; Division of Integrative Biosciences and Biotechnology, POSTECH, South Korea

CONTENTS

INTRODUCTION

Biomaterials are defined as any material, natural or man-made, that consists of whole or part of a living structure or a biomedical device that performs, augments, or replaces a function that has been lost through disease or injury [1].

In this regard, biomaterials are critical to tissue engineering, which combines scaffold or matrix with living cells and/or biologically active molecules to promote the repair and/or regeneration of tissues. As a result, tissue engineering treats biomaterials profoundly in the name of the scaffold. The main function of the scaffold is to direct the growth, migration, and differentiation of cells within its porous structure. Thus, the scaffolding materials should provide a suitable matrix for cell attachment, proliferation, migration, and differentiation functions. An ideal scaffold material should support cell attachment, promote cell proliferation, induce cellular response, and possess mechanical strength and dimensional stability, biodegradability, a high degree of processability, a high order of porosity, a high surface to volume ratio, and sterilizability [2].

Three-dimensional (3D) scaffold fabrication is a crucial issue in tissue engineering because it is called upon to satisfy various and tough requirements such as high-porosity, interconnectivity, sufficient or well-matched mechanical strength and stiffness, cell-friendly modified surface properties, nontoxicity, ease of fabrication and handling, and geometrical fidelity. All of these requirements are coupled with high complexity. In this regard, various factors that determine the characteristics of the scaffold must be treated in depth. Indeed, the proper design or selection of materials and methods to satisfy the requirements from target tissue can lessen the pressure of satisfying all requirements. Thus, an appropriate understanding of the materials, the methods, and the 3D scaffold itself is required.

3D porous scaffolds can be engineered from natural and synthetic materials and, in some cases, composites with ceramic materials. Accordingly, there are various candidates for materials and processing methods. Moreover, the range of tissues targeted in this field is expanding. Recent developments in biofabrication technology have created great expectation and enormous possibilities. Current biofabrication strategies offer material flexibility, with controlled architecture having favorable surface chemistry and predetermined degradation rates of the scaffolds [3]. In some tissue engineering initiatives, spatial but precise distribution of more than one type of cells is needed. These requirements are well fulfilled by the advanced manufacturing techniques [3]. It is thus envisioned that these attempts will enable the regeneration of not only single or multiple tissues but also complex organs like kidneys or livers for clinical use.

In this chapter, the broad range of materials used for the fabrication of scaffolds, the strategies applied for scaffold manufacturing, and the basics of various 3D tissue scaffolds are summarized.

2.1 **Materials for 3D tissue scaffolds**

Polymers are used widely as biomaterials for the fabrication of medical devices and tissue engineering scaffolds. Polymeric scaffolds are creating widespread interest among biomaterial scientists and tissue engineers. They offer distinct advantages in terms of biocompatibility, versatility of chemistry, and the biological properties that are important in tissue engineering and regenerative medicine [2]. Biomaterials are classified into several types with respect to their structural, chemical, and biological characteristics, such as naturally occurring polymers, synthetic biodegradable, and ceramics.

2.1.1 Biodegradable synthetic polymers

Biodegradable synthetic polymers are man-made materials that are highly useful in the biomedical field because their properties can be tailored to tissue engineering applications. Synthetic polymers show physicochemical and mechanical properties comparable to those of biological tissues. Generally, synthetic polymers are degraded by simple hydrolysis, and the degradation rate does not vary from host to host. The most extensively used synthetic polymers are poly(glycolic acid) (PGA), poly(lactic acid) (PLA), and their co-polymers poly(lactide-co-glycolide), polyanhydride, poly(propylene fumarate), polycaprolactone (PCL), polyethylene glycol (PEG), and polyurethane.

Poly(α-hydroxy ester)s are used as biodegradable scaffold materials due to their good biocompatibility, controlled biodegradability, and relatively good processability. These polymers degrade by nonspecific hydrolytic scission of their ester bonds, producing glycolic acid. PLA degrades more slowly than PGA due to its hydrophobic characteristic, and PLA−PGA co-polymers have no linear relationship with the ratio of PLA to PGA, although it is critical because the PLA hydrophobic characteristic reduces the rate of backbone hydrolysis, whereas PGA has a highly crystalline and crystallinity structure that leads to rapid loss in PLA−PGA co-polymers. PGA, PLA, and PLGA scaffolds are applied to the regeneration of all kinds of tissues, including skin, cartilage, blood vessels, nerves, and liver and other tissues [4].

Polyanhydride polymers are highly reactive and unstable synthetic biomaterials that are synthesized by the reaction of diacids with anhydride to form an acetal anhydride prepolymer. They are tailored to react with imides to increase the physical and mechanical properties resulting in polyanhydride-co-imides for applications in hard tissue engineering. These polymers possess excellent in vivo biocompatibility. The degradation mechanism of polyanhydride polymers is surface erosion, and their degradation rate is highly predictable compared to that of poly(α-hydroxy ester), which degrades by bulk erosion [5].

Polycaprolactone is a semicrystalline powder with a low melting point (59° to 64°C), which can form blends readily with other polymers. PCL is nontoxic and biocompatible and degrades through hydrolytic scission with resistance to rapid hydrolysis. PCL material has a significantly slower degradation rate; its complete degradation can take as long as 24 months. Thus, it is co-polymerized with other materials to generate the desired degradation properties. Due to its excellent biocompatibility, PCL has also been investigated extensively as a scaffold for tissue engineering [6].

Poly(propylene fumarate) and its copolymers are biodegradable unsaturated linear polyesters. The degradation mechanism of these polymers is hydrolytic chain scission, similar to that of poly(α-hydroxy ester). The degradation behavior and mechanical strength of these polymers are controlled by crosslinking of vinyl monomers with the unsaturated double bonds. The physical properties of poly(propylene fumarate) are enhanced by composite with calcium phosphate-degradable bioceramics for application as a bone substitute material [7].

Polyethylene oxide (PEO) possesses excellent biocompatibility and is used widely in biomedical applications. PEO hydrogels can be prepared by crosslinking reactions such as gamma ray, electron beam irradiation, or chemical reaction. The hydrogels are used in tissue engineering and drug delivery. Biodegradable PEO hydrogels can be synthesized by block copolymerization with PGA or PLA degradable units. These biodegradable hydrogels are very useful for injectable cell-loading scaffolds encapsulating chondrocyte cells, which degrade in vivo, and for neocartilage tissue formation [8].

Polyethylene glycol (PEG) is a hydrophilic synthetic polymer that restricts and controls the attachment of cells and proteins on scaffolds. PEG, being hydrophilic, prevents the adherence of other proteins to the scaffold surface and thereby minimizes any adverse immune response. By using PEG in a co-polymer, researchers can control the cell attachment characteristics of the scaffolds and enhance the biocompatibility of the co-polymer PEG [9].

Polyvinyl alcohol (PVA) is a biocompatible synthetic polymer that can swell to hold a large amount of water. It possesses reactive pendant alcohol groups and can be modified easily using physical or chemical crosslinking. It can be easily transformed to produce hydrogel, the properties of which can be tailored. PVA hydrogels have applications in cartilage regeneration, breast augmentation, and diaphragm replacement [10]. The limitation of using PVA scaffolds is their incomplete degradation.

Polyurethane is one of the most widely used polymeric biomaterials in biomedical fields due to its unique physical properties such as durability, elasticity, elastomer-like characteristics, fatigue resistance, compliance, and tolerance. Polyurethane is degradable by hydrolysis, oxidation, and thermal and enzymatic means. Toxicity, the induction of foreign-body reactions, and antibody formation have not been observed [11].

2.1.2 Natural polymers

Natural polymers are materials isolated from natural sources; these are used widely for clinical applications. Natural materials, owing to their biological recognition, have better interactions with cells, which allow them to improve the cellular interaction and performance of biological systems. Biocompatible natural biomaterials show minimal inflammatory response at the implantation site. It is very important to design appropriate materials for scaffold fabrication to mimic the natural ECMs in terms of bioactivity, mechanical properties, and structures.

Natural polymers can be classified as proteins (collagen, gelatin, silk, fibrin, etc.) and polysaccharides (alginates, agarose, chitosan, glycosaminoglycans, etc.).

Agarose is a polysaccharide consisting of a galactose-based backbone that is extracted from seaweeds, and it is used commonly as a cell culture medium. One of the attractive properties of agarose is that its stiffness can be altered, allowing for tuning of the mechanical properties of the scaffold. It can be used to form hydrogels, which gives the seeded cells a more uniform distribution throughout scaffolds. Agarose scaffolds degrade relatively slowly, similar to alginate. Agarose exhibits a temperature-sensitive solubility in water for encapsulating cells, and it has found widespread applications in tissue engineering [12].

Alginates are polysaccharides extracted from algae, consisting of two repeating monosaccharides: L-glucuronic acid and D-mannuronic acid. They can be formed as gels or beads, which are used for cartilage repair as a support to encapsulate cells in a scaffold. Through encapsulation, the chondrocyte phenotype during culture can be maintained; this allows cells that have been cultured in monolayers to be redifferentiated. Calcium alginate scaffolds do not degrade by hydrolytic cleavage; instead, they can be degraded by a chelating agent such as EDTA or by certain enzymes [13].

Chitosan is a deacetylated derivative of chitin, a naturally occurring polysaccharide extracted from crab shells, shrimps, and other shellfish or from a fungal fermentation process. Chitin and chitosan are semicrystalline polymers having a high degree of biocompatibility in vivo [14]. Chitosan can be prepared as a temperature-sensitive carrier material, and it is injectable as a fluid. It forms gels at body temperature and has the ability to deliver and interact with growth factors and adhesion proteins [15]. The degradation of chitosan is controlled by its residual acetyl content, and it can degrade rapidly in vivo, depending on the rate of polymer deacetylation.

Collagen, a fibrous protein, is used widely for tissue regeneration applications. There are more than 22 types of collagens found in the human body, among which Types I to IV are the most widely studied. Collagen provides cellular recognition for regulating cell attachment and proliferation [16]. It has relatively poor mechanical properties and less control over its biodegradation rate. Collagen can be processed into several types (sponge, felt, and fiber), gel, solution, filamentous material, tubular material (membrane and sponge), and composite matrix for tissue engineering applications [17].

Fibrin plays an important role in wound healing and thereby prevents blood loss during injury. Fibrin glue is formed from the mixture of fibrinogen and thrombin and allows them to solidify. It is often used as a carrier for cells and in conjunction with other scaffold materials. Fibrin gels can degrade by hydrolytic or enzymatic processes. Fibrin gels have applications as a tissue engineering scaffold matrix, especially for the repair of cartilage tissue [18].

Gelatin is derived from collagen and exhibits minimal immune response compared to collagen. Gelatin consists of a large number of glycine, proline, and 4-hydroxyproline residues. Gelatin is a denatured protein obtained by acid and alkaline processing of collagen. Due to its easy processability and gelation properties, gelatin can be engineered readily to fabricate sponges and injectable hydrogels. Gelatin matrices incorporating cell adhesion factors such as vitronectin, fibronectin, and RGD peptides enhance cell proliferation. Gelatin matrices have been used to treat and regenerate tissues like bone, cartilage, adipose tissue, and skin [19].

Glycosaminoglycans occur naturally as polysaccharide branches of a protein core to which they are covalently attached via a specific oligosaccharide linkage. Hyaluronan (HA) is an anionic polysaccharide that is used as a carrier for cells to regenerate various tissues. HA can be found abundantly within cartilaginous ECMs. It can be crosslinked to form scaffolds and seeded with chondrocytes and stem cells to induce chondrogenesis and osteogenesis on the scaffold [20]. Because it is injectable, it can be used in irregularly shaped defects and implanted with minimal invasion. Moreover, HA is a desirable biomaterial because it is not antigenic and elicits no inflammatory or foreign body reaction [21].

Silks are protein polymers spun by insects, such as silkworms and spiders. Silkworm silk consists of two types of proteins: fibroin and sericin. Silk fibroin has remarkable mechanical properties, biocompatibility, and controlled degradation rates, and it can be chemically modified to alter its surface properties. The conformational and eventual morphology of the silk fibroin chain is critical in the improved mechanical properties. Silk fibroin matrices are fabricated from regenerated silk solutions to form films, gels, 3D scaffolds, and so on. Silk fibroin scaffolds have been shown to support cell adhesion, proliferation, and differentiation in vitro and in vivo to engineer a range of tissues like bone, cartilage, tendon, and skin [22].

2.1.3 Biodegradable or resorbable ceramics

Bioceramics are biomaterials that are produced by sintering or melting inorganic raw materials to produce an amorphous or crystalline solid body, which can be used as an implant. Bioceramics used for tissue engineering are classified as nonresorbable (relatively inert), bioactive or surface active (semi-inert), and biodegradable or resorbable (noninert). Resorbable ceramics, as the name implies, degrade upon implantation in the host tissue and are replaced by endogenous tissue. The various bioresorbable ceramics used as biomaterials include calcium phosphates, tricalcium phosphates, hydroxyapatite, and bioglass. The rate of degradation of bioceramics is dependent on their composition and varies from material to material.

Calcium phosphates can be crystallized into salts such as hydroxyapatite or β-whitlockite, depending on the Ca:P ratio. They are very tissue-compatible and are used as bone substitutes as porous sponges or a solid block. Apatite formation with calcium phosphate is considered to be comparable to the mineral phase of bone and teeth. Calcium phosphate ceramics are tissue-compatible and are used as bone substitutes [23].

Tricalcium phosphate is a rapidly resorbable calcium phosphate ceramic that displays a resorption rate that is 10 to 20 times faster than hydroxyapatite. Porous tricalcium phosphate may stimulate local osteoblasts for new bone formation. Injectable calcium phosphate cement containing β-tricalcium phosphate, dibasic calcium phosphate, and tricalcium phosphate monoxide has been investigated for bone fractures [24].

Hydroxyapatite is chemically similar to the mineral component of bones and hard tissues in mammals. Hydroxyapatite is a naturally occurring mineral form of calcium apatite with the formula $Ca_5(PO_4)_3(OH)$. The ideal Ca:P ratio of hydroxyapatite is 10:6, and the calculated density is 3.219 g/cm^3. It is a bioactive material that supports bone growth and osteointegration when used in orthopedic, dental, and maxillofacial applications. Hydroxyapatite appears to form a direct chemical bond with hard tissues, and it exhibits excellent biocompatibility [25].

Bioactive glasses are surface reactive ceramics, which, upon implantation in the defect site, lead to strong osteointegration within the host tissue. Bioactive glass is a bioactive material that exhibits the capability to bond to both bone and soft tissue and stimulate bone growth [26].

2.2 Fabrication methods for 3D tissue scaffolds

To date, various fabrication techniques have been developed to produce three-dimensional (3D) scaffolds using synthetic and natural polymers. Each technique has its own strong and weak points. Therefore, a suitable fabrication method must be selected to satisfy the requirements related to the processing materials, geometric properties, and the characteristics of the target tissue. From the aspect of geometric properties of the fabricated product, the fabrication techniques can be classified into two categories: conventional and advanced technologies. The conventional methods cover processes of fabricating scaffolds with the help of traditional materials to make a porous structure. Therefore, a structure fabricated using a conventional process has a sponge- or formlike structure, with pores distributed uniformly, which lacks regularity in shape and interconnectivity. On the other hand, the advanced methods can fabricate the 3D scaffolds with control of geometric characteristics such as pore size, geometry, distribution, interconnectivity, and shape. As a result, the structure has well-defined geometry with orderly distributed pores with high regularity. In addition, there are several attractive approaches to conventional and advanced methods. In this regard, both types of methods still need to be investigated.

2.2.1 Conventional techniques for 3D scaffolds

Solvent-casting particulate-leaching has advantages in controlling pore size and porosity, which are determined by the diameter and amount of porogen added. This process consists of two steps: drying/solidifying and leaching. A polymer solution with porogen of a specific size range is prepared and then dried as a composite, which can be heated above the melting temperature of the polymer to adjust its crystallinity. To introduce porosity, the added porogen needs to be leached out into solvent. Typically, salt particles are used as the porogen. The structures produced by this process can have porosities of over 90 percent [27]. However, limitations in mechanical properties and problems from the residual porogen in the structure need to be addressed.

Freeze-drying can be applied to organic or glacial solvents to produce polymer solutions. With the use of organic solvent, water is added to emulsify the solution, and then the emulsion is frozen rapidly in a mold, followed by freeze-drying the water and solvent to yield a porous structure [28]. In the other case, the use of glacial acetic acid or benzene is called non-emulsion-based freeze-drying, which does not require the addition of water [29]. Also, a polymer that is insoluble in water can be dispersed in water to become a suspension for simple freeze-drying without solvent [30]. This technique is advantageous for the use of the mold, which can guide external shape. However, it has a limitation in pore sizes because of the emulsification or similar processes.

Melt molding has the advantage of excluding organic solvent from the fabrication process even though a high temperature is applied. This technique is based on conventional powdering and leaching processes. A mixture of polymer particles and water-soluble porogen is molded above glass transition or melting temperatures, followed by leaching to introduce porosity over the structure [31]. Because this technique uses solid materials without solvent, various solid materials such as ceramic particles or fibers, or bioactive molecules can be employed as additives.

Phase separation is used frequently in the production of porous structures of polymer. This method can be regarded as a more appropriate way to incorporate bioactive molecules like drugs with scaffold structures because toxic chemicals and high temperatures, which can degrade the bioactive molecules, are not part of the process. Briefly, the lowering of the temperature of a polymer solution with appropriate solvent results in phase separation, which is followed by the production of a two-phase solid by quenching. Next, sublimation of the solvent yields a porous polymer structure [32]. Bioactive molecules in particular can be loaded at the solution stage; a structure containing bioactive molecules is obtained after phase separation.

2.2.2 Advanced techniques for 3D scaffolds

Scaffold fabrication can incorporate solid free-form fabrication (SFF) technologies, which produces three-dimensional structures with complex geometry by stacking two-dimensional (2D) patterns in layer-by-layer [33,34]. SFF is an integrated technology of manufacturing, computer-aided engineering, and automation. Therefore, the convergence of scaffold fabrication and SFF can provide several benefits. First, various technologies such as computer-aided design (CAD), computer-aided manufacturing (CAM), and computed tomography (CT) are introduced to the fabrication of scaffolds. As a result, a scaffold with complex geometry or tight tolerance in shape, which could not be produced in a conventional fabrication environment, can be designed and fabricated with ease. Second, CAD, CAM, and CT are crucial parts of a computer-aided manufacturing (fabrication) environment, which means that the entire reconstruction of a new tissue or repair of a damaged tissue using a scaffold can be automated and become more well defined. Third, versatile SFF methods such as three-dimensional printing (3DP), stereo-lithography (SL), and fused deposition modeling (FDM) can be applied to scaffold fabrication. And finally, the fabrication resolution of SFF techniques can be improved for the fabrication of miniaturized parts. As a result, advanced scaffolds with complex geometry and micro-scale features can be fabricated by SFF.

In this regard, SFF methods can play a crucial role in producing scaffolds with customized external shapes and predefined and reproducible internal morphology, including pore size, porosity, and distribution. Moreover, the mass transport of oxygen and nutrients throughout the scaffold can be considered at the design stage through the geometry design parameter.

Stereolithography (SL) is selective polymerization of photocurable monomer resin using an ultraviolet (UV) laser beam. UV laser irradiation on the resin surface results in solidification of the irradiated surface. Therefore, UV beam scanning along a predefined path can generate a two-dimensional pattern, and thus the desired free-form structure can be produced through layer-by-layer solidification of 2D patterns [35,36]. Moreover, the scanning UV beam can be simplified by projection-based stereolithography. To improve the resolution of fabrication, micro-scale and nano-scale stereolithography were developed using elaborate optics and laser systems (Figure 2.1a). However, the application of this method to the fabrication of 3D scaffolds requires biodegradable but photocurable materials, the supply of which is limited. To avoid this limitation, an indirect SL method was developed [37,38]. In the indirect SL process, a sacrificial mold with an inverse porous shape of the desired scaffold is produced by SL. Then, a sacrificing mold process based on selective dissolution follows the injection of a polymer solution into the porous mold. This process is advantageous in combination with conventional techniques such as phase separation and particulate leaching. The selective laser sintering (SLS) method is similar to SL, but the mechanism is different. SLS typically uses a CO_2 laser to selectively heat and sinter various materials like ceramic and thermostable polymers just below their melting points [33].

Fused deposition modeling (FDM) fabricates 3D structures by stacking 2D patterns, which can be produced by extruding materials directly onto a plate [39,40]. In general, this process deposits a filament from liquefied material along a predefined path to construct 2D patterns. Filament length needs to be maintained to obtain consistent filament diameter, and thus the distance between the nozzle and deposited surface can be controlled (Figure 2.1b). This method does not require solvents and is relatively easier and more flexible

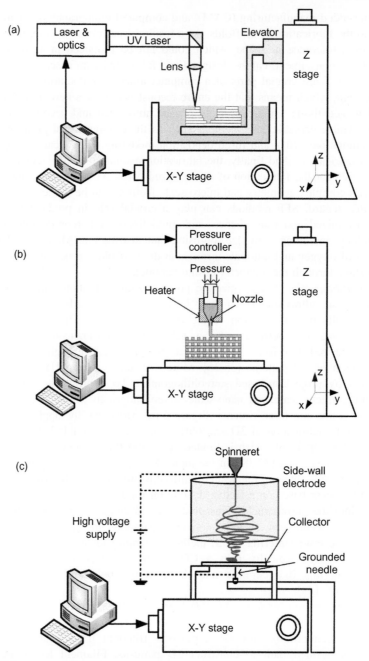

FIGURE 2.1

(a) Stereolithography apparatus. (b) Fusion deposition modeling apparatus.
(c) Direct-write electrospinning apparatus.

in terms of the handling and processing of materials compared to other SFF technologies.

Three-dimensional printing (3DP) employs the concept of conventional inkjet printing to eject a binder from a jet head [41], which moves in accordance to the cross-sectional data of scaffold structure, onto a polymer powder surface. The binder joins adjacent powder particles. Therefore, this process resembles laser sintering. The bound surface is lowered and a new layer of powder is deposited before repeating the process to produce another layer. 3DP is similar in process to FDM except for the head and powders, and thus 3DP can be used in FDM with polymer solution and in cell printing by ejecting hydrogel with cells [42]. Also, indirect approaches combining this process with conventional methods have been presented [43,44].

Direct-write electrospinning (DWES) is an advanced electrospinning process that can write a pattern of single nanofiber or nanofibers. Electrospinning, which is a conventional method for scaffold fabrication, uses an electric field force to draw nanofibers from a polymer solution [45,46]. However, the whipping motion of electrospun nanofibers makes the geometry difficult to control. As a result, the typical structure fabricated by conventional electrospinning is a mat of randomly deposited nanofibers, which is difficult to use as a 3D scaffold. However, randomly deposited nanofibers resemble the extracellular matrix of the human body, which can provide a human tissue-like environment to cells in culture. DWES with nanofiber focusing and scanning functionality was developed to fabricate a patterned nanofibrous mat with regular pores (resulting from patterns) in micro-scale and ECM-like random deposition of nanofiber, which can be used as a layer of a stacked nanofibrous 3D scaffold [47]. As a result, nanofibrous 3D scaffolds with well-defined pores and geometry can be fabricated (Figure 2.1c).

2.3 Application of 3D tissue scaffolds

Combining the efforts of numerous fields, tissue engineering is tackling the most significant and widespread clinical issues, and it brings hope for the rehabilitation of diseased or traumatized tissue/organs through utilizing the regeneration potential of the body's own cells seeded on biomaterial scaffolds. Thus, one of the key aspects of tissue engineering is the 3D scaffold, which provides support to the seeded or induced cells for fulfilling their mission of neo-tissue building. Recent advances in biofabrication methods like SFF have provided an opportunity to fabricate 3D scaffolds of controlled architecture, well-defined porosity, and biodegradation rates with various biomaterials. Further, decellularized ECM is a popular choice to fabricate biomimetic 3D scaffolds. To date, the highest rates of success have been achieved in the areas of skin, bladder, airway, bone, and cartilage, where tissue engineered constructs have been used successfully in patients [48]. In this section, the applications of different kinds of 3D scaffolds in TE are discussed.

2.3.1 Decellularized living tissues and organs

The extracellular matrix (ECM) of intact mammalian tissues has been used successfully as biological scaffold material in a variety of tissue engineering/regenerative medicine approaches, in both preclinical studies and clinical applications [49]. Decellularized extracellular matrices (dECM) have been shown to facilitate the constructive remodeling of many different tissues; however, the mechanisms by which functional tissue restoration is achieved are not completely understood. There is growing evidence to support the essential roles for both the structural and functional characteristics of dECM [49].

The goal of a decellularization protocol is to efficiently remove all cellular and nuclear material while minimizing any adverse effects on the composition, biological activity, and mechanical integrity of the remaining ECM [50]. The decellularization protocol varies from tissue to tissue, and it typically includes several processing steps like physical and chemical treatment, lyophilization, and sterilization, as described in detail elsewhere [49,50]. The dECM scaffolds are recellularized with tissue-specific cells for tissue/organ regeneration. The potential applications of decellularized matrix in tissue engineering have been demonstrated in a number of tissues, including bladder [51], artery [52], esophagus [53], skin [54], trachea [55], liver [56], and heart [57]. There have been a few in vivo studies of dECM for tissue regeneration. For example, Yoo et al. [51] investigated the possibility of using allogeneic bladder submucosa, either with or without cells, as a material for bladder augmentation. After 2 and 3 months of augmentation, a 99 percent increase in capacity was observed for bladders augmented with the allogeneic bladder submucosa. In another study, Nieponice et al. [53] reported that end-to-end anastomoses of the esophagus could be improved by reinforcement with an ECM scaffold without any systemic complications and with complete mucosal covering of the surgery site. Macchiarini et al. [55] used decellularized human donor trachea as a functional airway to replace the recipient's left main bronchus. The results of these studies have demonstrated the clinical potential of dECM as an effective treatment option for tissue or organ regeneration.

Whole-organ decellularization is a promising but challenging option for orthotopic organ transplantation. There have been a few attempts to use a decellularized whole organ for transplant. Uygun et al. [56] demonstrated an approach to generate transplantable liver grafts using decellularized liver matrix. In another study, Ott et al. [57] used decellularized hearts and reseeded with cardiac or endothelial cells; by day four, macroscopic contractions were observed, and by day eight, the construct generated pump function under physiological load and electrical stimulation (Figure 2.2).

2.3.2 Porous scaffolds with irregular pores

Porous 3D scaffolds play an imperative role in tissue engineering by controlling cell function and guiding new organ formation. A number of 3D porous scaffolds have been developed for tissue engineering of liver, bladder, nerve, skin, bone, cartilage, ligament, and so on [58]. Yannas et al. [59] led pioneering studies on collagen-glycosaminoglycan (CG) scaffolds to induce the regeneration of dermis of skin, sciatic nerve, and knee meniscus. Thereafter, a series of studies were conducted to fabricate porous scaffolds using various biomaterials [48,58,60]. Chemically crosslinked CG scaffold was also developed for bone regeneration, and the ability of these scaffolds to heal bone defects was also demonstrated in calvarial defects [61]. Biomimetic collagen-hydroxyapatite (CHA) scaffolds were developed based on the two primary constituents of bone to facilitate the repair of load-bearing regions [61]. Scaffolds shaped with poly (glycolic acid) (PGA), poly(lactic-co-glycolic acid) (PLGA), and poly-L-lactide (PLLA) fibers have been investigated for cell transplantation and regeneration of various tissues such as nerve, skin, esophagus, ligament, bladder, and cartilage [62]. These fibers are also used as reinforcing agents of polymeric materials for their application in orthopedic surgery [62]. PLLA scaffolds, with controlled microstructures, have been fabricated by a sugar sphere template leaching technique combined with a thermally induced phase separation (TIPS) method for bone repair [58]. A similar process using paraffin spheres was employed to fabricate 3D nanofibrous gelatin scaffolds and gelatin/apatite composites for bone TE applications [63].

2.3.3 Nanofibrous scaffolds

Electrospinning is the most promising and well-known technology used to produce nanofibrous scaffolds, and it has the ability to control precisely a vast array of parameters to tailor the scaffold for specific biomedical applications. The developed constructs serve as biomimetic extracellular matrices and enhanced cell adhesion, differentiation, and tissue formation [17].

A diverse array of synthetic polymers has been used to fabricate electrospun scaffolds for tissue engineering applications [64]. For bone tissue engineering, the osteogenic differentiation of human mesenchymal stem cells (hMSCs) has been evaluated on nanofibrous PLLA scaffolds in osteogenic media. It was also shown that hMSCs cultured on nanofibrous PLLA scaffolds in chondrogenic media for 6 weeks differentiated into chondrocytes. Electrospun PLLAs with polypropylene molds have been used to create 3D constructs for cartilage engineering applications. PCL nanofibers have been shown to support in vitro chondrogenesis of MSCs in the presence of TGF-β1. In another study, these nanofibers were demonstrated to support the growth and function of cardiomyocte cells. Aligned electrospun fibers are used commonly for nerve regeneration, since they have been shown to both guide and increase neurite outgrowth [64]. Electrospun collagen nanofibers closely resemble the natural dimensions and architecture of collagen fibrils of ECM. Vascular scaffolds composed of PCL and collagen have been prepared by electrospinning and provide a favorable environment for the growth of vascular cells [65].

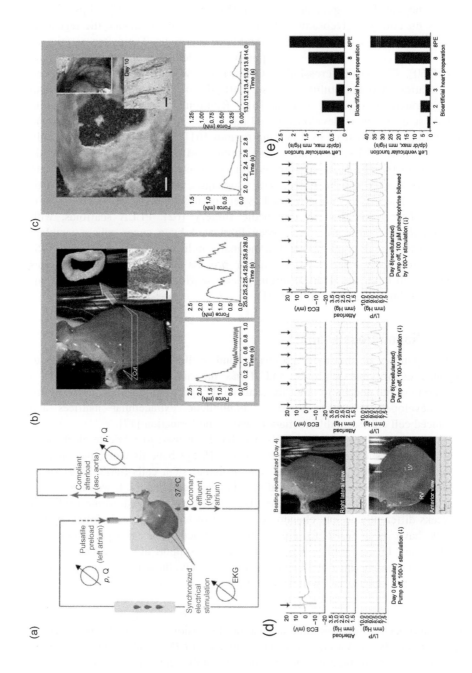

FIGURE 2.2

(a) Working heart bioreactor showing cannulation of left atrium and ascending (asc.) aorta. The heart is exposed to physiological preload, afterload, and intraventricular pressure and is electrically stimulated at 5–20 V. Oxygenated medium containing serum and antibiotics enters through the left atrium and exits through the aortic valve. Pulsatile distention of the LV and a compliance loop attached to the ascending aorta provide physiological coronary perfusion and afterload. Coronary perfusate (effluent) exits through the right atrium. (b) (Top) Recellularized whole rat heart at day 4 of perfusion culture in a working heart bioreactor. (Upper insert) Cross-sectional ring harvested for functional analysis (day 8); (lower insert) Masson's trichrome staining of a ring thin section showing cells throughout the thickness of the wall. Scale bar, 100 m. (Bottom) Force generation in left ventricular rings after 1-Hz (left) and 2-Hz (right) electrical stimulation. (c) (Top) Recellularized rat heart rings cultured for up to 10 d without perfusion. Scale bar, 250 m. (Upper insert) Microscopic view of cross-sectional ring showing rhythmic contractions at day 9; (lower insert) Masson's trichrome staining of ring thin section harvested for force generation studies after 10 d in vitro (scale bar, 50 m). (Bottom) Force generation in nonperfused rings at day 10, after 1-Hz (left) or 2-Hz (right) electrical stimulation. (d) (Left) Representative functional assessment tracing of decellularized whole heart construct paced in a working heart bioreactor preparation at day 0. Real-time tracings of ECG, aortic pressure (afterload), and left ventricular pressure (LVP) are shown. (Center) Paced recellularized heart construct on culture day 4 with pump turned off. Right lateral view (top) and anterior view (bottom) show physiological landmarks, including RV and LV. Tracing shows quantification of a region of movement in the beating preparation. (Right) Tracings of ECG (red), aortic pressure (afterload), and LVP of the paced construct on day 8 after recellularization and on day 8 after stimulation with physiological (50–100 M) doses of phenylephrine. (e) Summary of day 8 function in recellularized working heart preparation. Maximal developed LVP and dP/dt max obtained by day 8 in working heart bioreactor preparations.

Source: Reproduced with permission from [57], Nature Medicine, Nature Publishing Group.

Although the electrospun scaffolds provide favorable cellular interaction, cellular migration within 3D electrospun scaffolds has been limited, largely because of their inherent small pore sizes. Furthermore, the porosity of these scaffolds cannot be controlled. Fortunately, there are some new and emerging approaches to produce advanced electrospun 3D scaffolds for tissue regeneration applications where larger pores or higher porosity is needed [66]. Jeong and colleagues prepared 3D lattice-patterned nanofibrous scaffolds using a direct-write electrospinning system, where pore size and shape within the construct can be controlled precisely [47,67]. It is anticipated that these scaffolds will eventually improve cellular migration into the core and aid in 3D tissue formation (Figure 2.3).

2.3.4 Solid free-form (sff) scaffolds

SFF is an advanced manufacturing technique by which complex 3D structures can be manufactured by selectively stacking layers of 2D patterns that represent the cross section of the 3D structure. Generally, SFF includes various fabrication strategies like SL, FDM, and selective laser sintering (SLS) for the fabrication of 3D scaffolds, as reviewed by Seol et al. [33].

SFF techniques have been considered mainly for applications in hard tissue engineering, although recent advances in this field have also witnessed their application for other soft tissue engineering. Through these techniques,

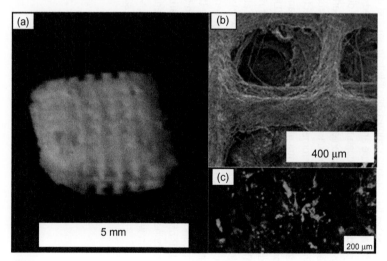

FIGURE 2.3

SEM images of (a) whole scaffold, (b) pores within scaffold, and (c) cellular attachment and migration 3D LPN scaffold after 14 days at cross-section view. F-actin and nucleus of the cells were stained by phalloidin-FITC (green) DAPI (blue).

Source: Reproduced with permission from [67].

scaffolds with a resolution of several tenths of a micron have been fabricated successfully using various biomaterials, such as polymer and blended polymer/bioceramic [40]. Cho et al. fabricated SFF scaffolds with poly(propylene fumarate) (PPF) and a bone morphogenetic protein 2 (BMP-2) mixture, which showed promise for bone tissue regeneration (Figure 2.4) [68]; they also fabricated an HA/PLGA conjugate scaffold with an intact BMP-2/PEG complex for bone tissue regeneration [33].

SLS has been used to fabricate ceramic, polymeric, and composite scaffolds, mostly for bone tissue engineering applications. SLS was used to fabricate scaffolds from PCL in order to successfully construct prototypes of mini-pigs' mandibular condyle scaffolds [63]. Biocomposite blends of PCL with different percentages of HA were subjected to an evaluation of their suitability for fabrication via SLS, which can also replicate the desired anatomy precisely [69]. SL was employed to fabricate micropatterned scaffolds for both bone tissue engineering applications using a blend of diethyl fumarate and poly(propylene fumarate) [33]. SL has also been employed for soft tissue engineering applications using photolabile hydrogels; however, the viability and uniformity of the encapsulated cells were an issue in those structures. Chan et al. modified the SL process from its conventional top-down version to a bottom-up approach [63]. FDM has been used to fabricate a number of scaffolds with highly interconnecting and controllable pore structure, mainly for bone tissue engineering

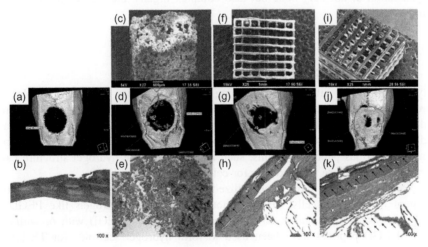

FIGURE 2.4

Rat cranial bone regeneration at 11 weeks after implantation: (a, b) micro-CT and H&E stained images of negative control; (c–e) SEM, micro-CT, H&E stained images of BMP-2 unloaded traditional scaffold; (f–h) SEM, micro-CT, H&E stained images of BMP-2 unloaded SFF scaffold; (i–k) SEM, micro-CT, H&E stained images of BMP-2 loaded SFF scaffold (black arrows, regenerated bone; blue arrows, scaffold).

Source: Reproduced with permission from [68].

applications using synthetic polymers such as PCL and PLGA and their composites with ceramics [63].

Indirect SFF techniques have been used to fabricate microgrooved molds from polydimethylsiloxane (PDMS) and chemically coated with a temperature-responsive polymer, poly(N-isopropylacrylamide) (PNIPAAm), to generate size- and shape-controlled tissue constructs [63]. Interestingly, one can generate the entire micromold from PNIPAAm, and 3D microtissues could be created from patterned cells sequentially encapsulated in agarose hydrogels using PNIPAAm-patterned molds [63]. Furthermore, multicomponent hydrogels have been generated using both NIH-3T3 cells and human umbilical vein endothelial cells (HUVECs) or HepG2 cells and HUVECs using the same patterns [63].

The dispensing-based SFF technology enables direct writing of cells. Initially, commercial inkjet printing heads were used to print cells in hydrogels; however, their exact 3D positioning or mechanical support could not be controlled. Interestingly, multiple nozzle deposition systems enable the fabrication of hybrid structures consisting of biomaterials/cell/growth factors. These hybrid structures can be considered as the ultimate prepackaged tissue replacements, with exact 3D positioning [33]. PEG is one of the most hydrophilic polymers, and its acrylated form (PEG-diacrylate or PEGDA) has been used extensively for fabricating cell-laden micropatterned structures [63]. SFF technology has created a huge expectation and vast possibilities. It is envisioned that these attempts will empower the regeneration of not only bone but also complex organs like kidney and liver for clinical use.

2.3.5 Hydrogel scaffolds

Hydrogels are an attractive scaffolding material because they are capable of providing bulk and mechanical constitution to a tissue construct, whether cells are adhered to or suspended within the 3D gel framework [70]. When direct cellular adhesion to the gel is favored over suspension within the scaffold, the incorporation of various peptide domains, such as RGD, into the hydrogel structure can dramatically increase the tendency for cellular attachment [70].

Various strategies have been utilized for the gelation of hydrogels with or without incorporated cells. Mainly, photopolymerizable hydrogels like PEGDA and poly(propylene fumarate-co-ethylene glycol) (P[PF-co-EG]) with or without grafted RGD peptides have been used to form hydrogel scaffolds for TE [70]. Hydrogels have also been used as barriers to improve the healing response following tissue injury to prevent restenosis or thrombosis due to postoperative adhesion formation [70]. Cell transplantation can be achieved with hydrogels because they can provide immunoisolation while still allowing oxygen, nutrients, and metabolic products to diffuse easily into the hydrogel. For the development of a bioartificial endocrine pancreas, photopolymerized PEGDA hydrogel has been employed to transplant islets of Langerhans [71].

Naturally derived hydrogels like hyaluronic acid (HA), fibrin, alginate, chitosan, and collagen are generally considered to have an edge over synthetic biomaterials where biocompatibility is concerned, because natural gels may offer better chemical and morphological cues to cells. These hydrogels have been used for various applications, including wound healing, soft tissue augmentation, ophthalmic surgery, and treatment of osteoarthritis [70]. Langer et al. [72] used HA to make micromolds for cell encapsulation, where encapsulated cells could later be recovered by enzymatic degradation. Composites of collagen with HA have also been prepared in order to stabilize chondrocyte phenotype and enhance proteoglycan synthesis [70]. In another study, collagen−alginate and collagen−hyaluronan composites were used to restore the appropriate shape and pliability to scarred vocal folds [70]. Collagen gels are also being used as nerve guidance materials in the spinal cord and peripheral nervous system [70].

The majority of tissues are vascularized, which provides them a conduit for nutrient exchange and the elimination of waste products by perfusion. Neovascularization is therefore an important consideration for most TE initiatives. Alginate, gelatin, HA, PHEMA, and PEG-based hydrogels loaded with vasculogenic growth factors have been shown to successfully induce microvessel growth following implantation [70]. Furthermore, in order to spatially control the growth of new vasculature, Golden and Tien [73] designed endothelial cell (EC)-seeded microfluidic channels in collagen and fibrin hydrogel scaffolds.

Other important hydrogels are those made from self-assembled peptides (SAPs). Zhang et al. [74] developed a class of SAPs where peptides self-assemble into beta sheets and subsequently form hydrogels. A variety of cell types can be encapsulated within these hydrogels and used for generating 3D environments for cell culture and tissue engineering applications. Recently, it has been demonstrated that the controlled assembly of self-assembled peptides, along with molecules such as HA domains, can be used to form strong membranes that are potentially useful for generating tissue engineered structures, such as blood vessels and membranes [70]. It is thus envisioned that future developments in this area will be beneficial for generating artificial tissues.

2.4 Future perspectives of biomaterials for 3D scaffolds

Numerous efforts have been made to develop suitable materials for biofabrication, with some successful and encouraging results. In the future, additional developments of materials for tissue fabrication are necessary to facilitate the progress of scaffold fabrication technology. Scaffold design for tissue regeneration is one of the key technologies used in tissue engineering. Although synthetic materials are suitable for biofabrication, it is important to explore additional biological materials in future studies. Recent biochemical and nano-biotechnological processes have introduced modified biomaterials, such as RGD-modified alginate hydrogel, and ECM proteins immobilized with growth factors and biofunctional peptide domains [70]. The design and precise control of several biochemical signals must be properly oriented using the combinatorial effects of the scaffold and biologically active molecules like genes, cytokines, and peptides. Indeed, exploratory research for more suitable and effective biological materials for use in bioprinting and biofabrication procedures is under way. It is thus expected that a reasonable development cycle between fabrication and material technology advancements will occur, which will facilitate bioprinting and biofabrication methods to realize the goal of functional manufactured human tissues and organs. There have been some successful attempts to regenerate functional tissues or organs in laboratories worldwide. However, the application of artificial functional tissue or organs in a clinical setting represents the ultimate accomplishment; this will likely be achieved in the near future.

CONCLUSION

This chapter outlines the different types of biodegradable materials (synthetic, natural, and ceramics) used for the development of scaffolds, as well as various conventional and advanced methods for 3D scaffold fabrication. In the latter part of the chapter, we discussed the application of the various forms of 3D tissue-specific scaffolds and future perspectives on biomaterials research in tissue regenerative therapy. The development of adequate biodegradable and biocompatible materials for each fabrication process is critical for engineering tissue scaffolds. Future developments in biofabrication will depend on the advanced design of novel scaffolding materials, precision-based scaffold fabrication systems, and enhanced studies of cell-matrix interactions, including optimal cell adhesion, proliferation, and matrix remodeling. With some of the recent innovative developments in this field, the future of tissue/organ regeneration looks bright. Such technologies can relieve the suffering of patients and bring smiles to their faces.

Acknowledgment

This work was supported by the National Research Foundation of Korea (NRF) grant funded by the Korean government (MEST) (No. 2011-0000412).

References

[1] Williams DF. Definition in biomaterials. Progress in biomedical engineering. Amsterdam: Elsevier; 1987;4:54.
[2] Lee HB, Khang G, Lee JH. Polymeric biomaterials. In: Park JB, Bronzino JD, editors. Biomaterials: principles and applications. Boca Raton, FL: CRC Press; 2003.

[3] Melchels FPW, Domingos MAN, Klein TJ, Malda J, Bartolo PJ, et al. Additive manufacturing of tissues and organs. Prog Polym Sci 2011. [In Press].

[4] Sun H, Qu Z, Guo Y, Zang G, Yang B. In vitro and in vivo effects of rat kidney vascular endothelial cells on osteogenesis of rat bone marrow mesenchymal stem cells growing on polylactide-glycolic acid (PLGA) scaffolds. Biomed Eng Online 2007;6:41.

[5] Sabir M, Xu X, Li L. A review on biodegradable polymeric materials for bone tissue engineering applications. J Mater Sci 2009;44:5713−24.

[6] Fuchs S, Ghanaati S, Orth C, Barbeck M, Kolbe M, et al. Contribution of outgrowth endothelial cells from human peripheral blood on in-vivo vascularization of bone tissue engineered constructs based on starch polycaprolactone scaffolds. Biomaterials 2009;30:526−34.

[7] Peter SJ, Miller MJ, Yaszemski MJ, Mikos AG. Poly(propylene fumarate). In: Domb AJ, Kost J, Wiseman DM, editors. Handbook of biodegradable polymers. Amsterdam: Harwood Academic; 1997: 87−97.

[8] Sanabria-Delong N, Crosby AJ, Tew GN. Photo-cross-linked PLA-PEO-PLA hydrogels from self-assembled physical networks: mechanical properties and influence of assumed constitutive relationships. Biomacromolecules 2008;9:2784−91.

[9] Lin CC, Anseth K. PEG hydrogels for the controlled release of biomolecules in regenerative medicine. Pharm Res 2009;26:631−43.

[10] Bryant SJ, Davis-Arehart KA, Luo N, Shoemaker RK, Arthur JA, et al. Synthesis and characterization of photopolymerised multifunctional hydrogels: water-soluble poly(vinyl alcohol) and chondroitin sulfate macromers for chondrocyte encapsulation. Macromolecules 2004;37.

[11] Yoshii T, Hafeman AE, Nyman JS, Esparza JM, Shinomiya K, et al. A sustained release of lovastatin from biodegradable, elastomeric polyurethane scaffolds for enhanced bone regeneration. Tissue Eng 2010;16:2369−79.

[12] Finger AR, Sargent CY, Dulaney KO, Bernacki SH, Loboa EG. Differential effects on messenger ribonucleic acid expression by bone marrow-derived human mesenchymal stem cells seeded in agarose constructs due to ramped and steady applications of cyclic hydrostatic pressure. Tissue Eng 2007;13:1151−8.

[13] Ashton RS, Banerjee A, Punyani S, Schaffer DV, Kane RS. Scaffolds based on degradable alginate hydrogels and poly(lactide-co-glycolide) microspheres for stem cell culture. Biomaterials 2007;28:5518−25.

[14] Nolan K, Millet Y, Ricordi C, Stabler CL. Tissue engineering and biomaterials in regenerative medicine. Cell Transplant 2008;17:241−3.

[15] Shi CM, Zhu Y, Ran XZ, Wang M, Su YP, et al. Therapeutic potential of chitosan and its derivatives in regenerative medicine. J Surg Res 2006;133:185−92.

[16] Kadler KE, Baldock C, Bella J, Boot-Handford RP. Collagens at a glance. J Cell Sci 2007;120:1955−8.

[17] Nair LS, Laurencin CT. Biodegradable polymers as biomaterials. Prog Polym Sci 2007;32.

[18] Zhao H, Ma L, Zhou J, Mao Z, Gao C, et al. Fabrication and physical and biological properties of fibrin gel derived from human plasma. Biomed Mater Eng 2008;3:15001.

[19] Lien S-M, Ko L-Y, Huang T-J. Effect of pore size on ECM secretion and cell growth in gelatin scaffold for articular cartilage tissue engineering. Acta Biomater 2009;6:670−9.

[20] Kogan G, Soltes L, Stern R, Gemeiner P. Hyaluronic acid: a natural biopolymer with a broad range of biomedical and industrial applications. Biotechnol Lett 2007;29:17−25.

[21] Gerecht S, Burdick J, Ferreira L, Townsend S, Langer R, et al. Hyaluronic acid hydrogel for controlled self-renewal and differentiation of human embryonic stem cells. Proc Natl Acad Sci USA 2007;104:11298.

[22] Rockwood DN, Preda RC, Yucei T, Wang X, Lovett ML, et al. Materials fabrication from Bombyx mori silk fibroin. Nat Protoc 2011;6:1612−31.

[23] Mastrogiacomo M, Scalione S, Marinetti R, Dolcini L, Beltrame F, et al. Role of scaffold internal structure on in vivo bone formation in macroporous calcium phosphate ceramics. Biomaterials 2006;27:3230−7.

[24] Zhang Y, Zhang M. Cell growth and function on calcium phosphate reinforced chitosan scaffolds. J Mater Sci Mater Med 2004;15.

[25] Okuda T, Ioku K, Yonezawa I, Minagi H, Gonda Y, et al. The slow resorption with replacement by bone of a hydrothermally synthesized pure calcium-deficient hydroxyapatite. Biomaterials 2008;29:2719−28.

[26] Hench LL, Polak JM. Third-generation biomedical materials. Science 2002;295:1014−7.

[27] Mikos AG, Lyman MD, Freed LE, Langer R. Wetting of poly (L-lactic acid) and poly (DL-lactic-co-glycolic acid) foams for tissue culture. Biomaterials 1994;15:55−8.

[28] Whang K, Thomas C, Healy K, Nuber G. A novel method to fabricate bioabsorbable scaffolds. Polymer (Guildf) 1995;36:837−42.

[29] Hsu YY, Gresser JD, Trantolo DJ, Lyons CM, Gangadharam PRJ, et al. Effect of polymer foam morphology and density on kinetics of in vitro controlled release of isoniazid from compressed foam matrices. J Biomed Mater Res 1997;35:107−16.

[30] Yannas I, Burke J, Gordon P, Huang C, Rubenstein R. Design of an artificial skin. II. Control of chemical composition. J Biomed Mater Res 1980;14:107−32.

[31] Thomson RC, Yaszemski MJ, Powers JM, Mikos AG. Fabrication of biodegradable polymer scaffolds to engineer trabecular bone. J Biomater Sci Polym Ed 1996;7:23−38.

[32] Lo H, Ponticiello M, Leong K. Fabrication of controlled release biodegradable foams by phase separation. Tissue Eng 1995;1:15−28.

[33] Seol Y-J, Kang T-Y, Cho D-W. Solid freeform fabrication technology applied to tissue engineering with various biomaterials. Soft Matter 2012;8:1730−5.

[34] Hutmacher DW, Sittinger M, Risbud MV. Scaffold-based tissue engineering: rationale for computer-aided design and solid free-form fabrication systems. Trends Biotech 2004;22:354−62.

[35] Hollister SJ. Porous scaffold design for tissue engineering. Nat Mater 2005;4:518−24.

[36] Lee JW, Lan PX, Kim B, Lim G, Cho DW. Fabrication and characteristic analysis of a poly (propylene fumarate) scaffold using micro-stereolithography technology. J Biomed Mater Res B 2008;87:1−9.

[37] Kang HW, Seol YJ, Cho DW. Development of an indirect solid freeform fabrication process based on microstereolithography for 3D porous scaffolds. J Micromechanic Microengineer 2009;19:015011.

[38] Kang HW, Cho DW. Development of an indirect stereolithography technology for scaffold fabrication with a wide range of biomaterial selectivity. Tissue Eng 2012;18:719−29.

[39] Zein I, Hutmacher DW, Tan KC, Teoh SH. Fused deposition modeling of novel scaffold architectures for tissue engineering applications. Biomaterials 2002;23:1169−85.

[40] Kim JY, Jin G-Z, Park IS, Kim J-N, Chun SY, et al. Evaluation of solid free-form fabrication-based scaffolds seeded with osteoblasts and human umbilical vein endothelial cells for use in vivo osteogenesis. Tissue Eng Part A 2010;16:2229−36.

[41] Sachs E, Cima M, Williams P, Brancazio D, Cornie J. Three-dimensional printing: rapid tooling and prototypes directly from a CAD model. J Eng Ind-T ASME (USA) 1992;114:481−8.

[42] Mironov V, Boland T, Trusk T, Forgacs G, Markwald RR. Organ printing: computer-aided jet-based 3D tissue engineering. Trends Biotech 2003;21:157−61.

[43] Taboas J, Maddox R, Krebsbach P, Hollister S. Indirect solid free form fabrication of local and global porous, biomimetic and composite 3D polymer-ceramic scaffolds. Biomaterials 2003;24:181−94.

[44] Lee M, Dunn JCY, Wu BM. Scaffold fabrication by indirect three-dimensional printing. Biomaterials 2005;26:4281−9.

[45] Li WJ, Laurencin CT, Caterson EJ, Tuan RS, Ko FK. Electrospun nanofibrous structure: a novel scaffold for tissue engineering. J Biomed Mater Res 2002;60:613−21.

[46] Yoshimoto H, Shin Y, Terai H, Vacanti J. A biodegradable nanofiber scaffold by electrospinning and its potential for bone tissue engineering. Biomaterials 2003;24:2077−82.

[47] Lee J, Lee SY, Jang J, Jeong YH, Cho D-W. Fabrication of patterned nanofibrous mats using a direct-write electrospinning. Langmuir 2012;28:7267−75.

[48] O'Brien FJ. Biomaterials & scaffolds for tissue engineering. Mater Today 2011;14:88−95.

[49] Badylak SF, Freytes DO, Gilbert TW. Extracellular matrix as a biological scaffold material: structure and function. Acta Biomater 2009;5:1−13.

[50] Badylak SF, Taylor D, Uygun K. Whole-organ tissue engineering: decellularization and recellularization of three-dimensional matrix scaffolds. Annu Rev Biomed Eng 2011;13:27−53.

[51] Yoo JJ, Meng J, Oberpenning F, Atala A. Bladder augmentation using allogenic bladder submucosa seeded with cells. Urology 1998;51:221−5.

[52] Dahl SL, Koh J, Prabhakar V, Niklason LE. Decellularized native and engineered arterial scaffolds for transplantation. Cell Transplant 2003;12:659−66.

[53] Nieponice A, Gilbert TW, Badylak SF. Reinforcement of esophageal anastomoses with an extracellular matrix scaffold in a canine model. Ann Thorac Surg 2006;82:2050−8.

[54] Schechner JS, Crane SK, Wang F, Szeglin AM, Tellides G, et al. Engraftment of a vascularized human skin equivalent. FASEB J 2003;17:2250−6.

[55] Macchiarini P, Jungebluth P, Go T, Asnaghi MA, Rees LE, et al. Clinical transplantation of a tissue-engineered airway. Lancet 2008;372:2023−30.

[56] Uygun BE, Soto-Gutierrez A, Yagi H, Izamis M-L, Guzzardi MA, et al. Organ reengineering through development of a transplantable recellularized liver graft using decellularized liver matrix. Nat Med 2010;16:814−21.

[57] Ott HC, Matthiesen TS, Goh S-K, Black LD, Kren SM, et al. Perfusion-decellularized matrix: using nature's platform to engineer a bioartificial heart. Nat Med 2008;14:213−21.

[58] Lee J, Cuddihy MJ, Kotov NA. Three-dimensional cell culture matrices: state of the art. Tissue Eng: Part B 2008;14:61−86.

[59] Yannas IV, Burke JF, Orgill DP, Skrabut EM. Wound tissue can utilize a polymeric template to synthesize a functional extension of skin. Science 1982;215:174−6.

[60] Ikada Y. Review: Challenges in tissue engineering. J R Soc Interface 2006;3:589−601.

[61] Lyons FG, Al-Munajjed AA, Kieran SM, Toner ME, Murphy CM, et al. The healing of bony defects by cell-free collagen-based scaffolds compared to stem cell-seeded tissue engineered constructs. Biomaterials 2010;31:9232−43.

[62] Armentano I, Dottori M, Fortunati E, Mattioli S, Kenny JM. Biodegradable polymer matrix nanocomposites for tissue engineering: a review. Polym Degrad Stabil 2010;95:2126−46.

[63] Zorlutuna P, Jeong JH, Kong H, Bashir R. Tissue engineering: stereolithography-based hydrogel microenvironments to examine cellular interactions. Adv Funct Mater 2011;21:3597.

[64] Nisbet DR, Forsythe JS, Shen W, Finkelstein DI, Horne MK. Review paper: a review of the cellular response on electrospun nanofibers for tissue engineering. J Biomater Appl 2009;24:7−29.

[65] Burger C, Hsiao BS, Chu B. Nanofibrous materials and their applications. Ann Rev Mater Res 2006;36:333−68.

[66] Zhong S, Zhang Y, Lim CT. Fabrication of large pores in electrospun nanofibrous scaffolds for cellular infiltration: a review. Tissue Eng Part B 2012;18:77−87.

[67] Lee J, Jang J, Oh H, Jeong YH, Cho D-W. Fabrication of three-dimensional lattice patterned nanofibrous scaffold using a direct-write electrospinning apparatus. Mater Lett 2013;93:397−400.

[68] Lee JW, Kang KS, Lee SH, Kim J-Y, Lee B-K, et al. Bone regeneration using a microstereolithography-produced customized poly(propylene fumarate)/diethyl fumarate photopolymer 3D scaffold incorporating BMP-2 loaded PLGA microspheres. Biomaterials 2011;32:744−52.

[69] Wiria FE, Leong KF, Chua CK, Liu Y. Poly-ε-caprolactone/hydroxyapatite for tissue engineering scaffold fabrication via selective laser sintering. Acta Biomater 2007;3:1−12.

[70] Slaughter BV, Khurshid SS, Fisher OZ, Khademhosseini A, Peppas NA. Hydrogels in regenerative medicine. Advan Mater 2009;21:3307−29.

[71] Cruise GM, Hegre OD, Lamberti FV, Hager SR, Hill R, et al. In vitro and in vivo performance of porcine islets encapsulated in interfacially photopolymerized poly (ethylene glycol) diacrylate membranes. Cell Transplant 1999;8:293−306.

[72] Yeha J, Lingb Y, Karpd JM, Gantze J, Chandawarkard A, et al. Micromolding of shape-controlled, harvestable cell-laden hydrogels. Biomaterials 2006;27:5391−8.

[73] Golden AP, Tien J. Fabrication of microfluidic hydrogels using molded gelatin as a sacrificial element. Lab Chip 2007;7:720−5.

[74] Zhang S. Fabrication of novel biomaterials through molecular self-assembly. Nat Biotech 2003;21:1171−8.

Projection Printing of Three-Dimensional Tissue Scaffolds with Tunable Poisson's Ratio

3

Pranav Soman, Shaochen Chen

Department of NanoEngineering, University of California, San Diego, La Jolla, CA

CONTENTS

INTRODUCTION

In the field of tissue engineering, a biomaterial scaffold's elastic properties are critical to its efficacy in regenerating tissue and reducing inflammatory responses, and they must be matched with the elastic properties of the native tissue it will replace [1,2]. The ability of a biomaterial scaffold to support and transmit cell and tissue forces can be quantitatively described by its elastic modulus and Poisson's ratio. Elastic modulus of the underlying substrate quantifies a scaffold's elastic behavior in the loading direction, while Poisson's ratio describes the degree to which the scaffold contracts/expands in the transverse direction. Elastic modulus is of vital importance in determining whether a scaffold demonstrates satisfactory mechanical integrity. So far, the elastic modulus of scaffolds has been tuned by modulating properties such as crosslink density and swelling ratio [3]. Tuning the Poisson's ratio, however, requires control over pore-interconnectivity and internal architecture of a scaffold, which can be quite difficult to achieve and

Biofabrication.

has been generally assumed to be between $+0.3$ and $+0.5$. In some applications, scaffolds having a negative Poisson's ratio (NPR) may be more suitable for emulating the behavior of native tissues and accommodating and transmitting forces to the host tissue site [4—11]. When axial forces are applied, a positive Poisson's ratio (PPR) scaffold contracts in the transverse direction, while an NPR scaffold would expand biaxially in both the axial and transverse directions simultaneously.

In nature, we come across several materials with an NPR, such as in crystal structures [12,13], carbon allotropes [14], foams [15—17], polymers, and laminates [18—21]. In an attempt to mimic nature, several researchers have developed polymers with NPR behavior or "auxetic polymers" by incorporating rib-containing pores. These pores modify the shape and deformation mechanisms of polymers, transforming their mechanical behavior to exhibit an auxetic property [15,17,19,22—26]. Several man-made polymers, incorporated with unit-cells or pores, demonstrate well-defined strain-dependent Poisson's ratio behaviors [15,27—29]. However, the resulting auxetic behavior in man-made polymer is "process dependent" and cannot be tuned according to a specific value of Poisson's ratio. For example, polyurethane foams annealed in a compressed state naturally reorganize their cellular microstructure and exhibit NPR behavior [15,22]. Conventional manufacturing processes such as annealing have little control over the reorganization of cellular microstructure comprising the foams, making it difficult to tune the Poisson's ratio according to a specific application. For a tissue engineering application, Poisson's ratio of a scaffold should be precisely tuned, both in magnitude and polarity, to match the properties of the target tissue being regenerated. Recently, we developed 3D scaffolds that exhibit negative, positive, and zero Poisson's ratio (ZPR) behavior in poly(ethylene glycol) diacrylate (PEGDA) biomaterials. Digital micromirror-assisted projection printing stereolithography (DMD-PP) was used to generate dynamic masks of several unit-cell models, which tuned the magnitude and polarity of the scaffold and accordingly exhibited different Poisson's ratio behavior. In this chapter, we discuss the DMD-PP platform and its versatility in fabricating tissue engineering scaffolds with tunable Poisson's ratio.

3.1 Digital micromirror-assisted projection printing stereolithography

Conventional fabrication methods used for manufacturing 3D scaffolds, such as freeze-drying [30], electrospinning [31], and gas foaming [32], do not allow precise control of the internal structural features and topology. Free-form fabrication techniques like digital micromirror device-based projection printing (DMD-PP), developed in our lab, are capable of fabricating complex 3D scaffolds with precise microarchitectures [33—35]. As shown in Figure 3.1a, the DMD-PP system consists of five parts: a UV light source, a computer for sliced image-flow generation

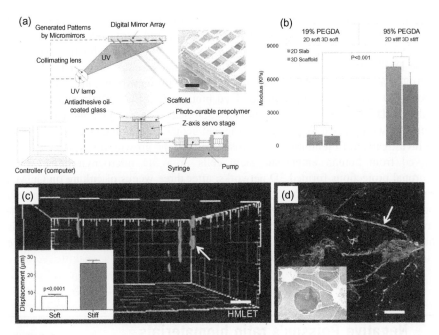

FIGURE 3.1

(a) Schematic of the digital micromirror device—based projection printing (DMD-PP) used for fabricating multilayer biomaterial scaffolds having complex and precise geometry. Inset shows an SEM image of a logpile PEGDA scaffold. (b) Mechanical properties of scaffolds having 2D slab and 3D logpile structures and fabricated using 19% (soft scaffold) and 95% (stiff scaffold) PEGDA (700 kDa molecular weight). Prepolymer concentration and 3D architecture can be used to tailor the mechanical properties of the PEGDA scaffolds. (c) Still images at 8 hours from reconstituted 3D confocal stacks of TWIST oncogene-modified breast epithelial cells (HMLET) on soft and stiff 3D PEG log pile scaffolds. Inset shows significant difference of total displacement of HMLET cells between soft and stiff 3D scaffolds. (d) GelMA scaffolds with neural stem cells after 3 to 4 weeks of culture. Tuj1 (Green)J Map2 (Red) and DAPI. Inset shows 3D neural connections on PEG scaffolds with hexagonal geometry. (Scale: 100 μm)

and system synchronization, a DMD chip for optical pattern generation, projection optics, and a stage for sample position control. A collimated UV beam is spatially modulated by the DMD chip to generate a pattern on a photosensitive prepolymer solution. Precise 3D microstructures can be fabricated using a layer-by-layer polymerization process. Recently, we used the DMD-PP system to fabricate 3D logpile structures that show distinct compressive modulus, called soft (19 percent PEGDA, ~0.9 MPa) and hard (95 percent PEGDA, ~7 MPa) scaffolds (Figure 3.1b). These "design-driven" hard and soft logpile scaffolds were used to investigate 3D migration properties of normal and TWIST-transformed cancerous breast epithelial cells (HMLET) [36]. Cancer cell migration in 3D yielded significant differences in displacement, velocity, and straightness as compared with 2D cell migration (Figure 3.1c). The complete control over the elastic modulus and internal geometry in the DMD-PP process allows investigation of concurrent effects of modulating multiple parameters of cell migration. The DMD-PP approach can also fabricate precise 3D structures in naturally derived biomaterials like gelatin or hyaluronic acid. DMD-PP was able to fabricate 3D scaffolds in gelatin-methacrylate (GelMA), with precise logpile and hexagonal geometries. Using GelMA scaffolds, we investigated differentiation of neurospheres and the growth of passaged neurons derived from human embryonic stem cells in a 3D microenvironment [37]. Neuronal projections formed 3D networks, after three weeks of culture on GelMA scaffolds and express neuronal-specific microtube-associated protein MAP2 (red) and Tuj (green) (Figure 3.1d). DMD-PP technology can be used to standardize culture conditions for cell-based drug and toxicity screenings using stem cells. In this chapter, we focus on another unique aspect of DMD-PP: the ability to tune Poisson's ratio in any photocurable polymer.

3.2 Negative Poisson's ratio biomaterials

A scaffold with an NPR or "auxetic" property would allow a biomaterial to expand or compress uniformly or nonuniformly in the axial and transverse directions. Recently, it has been demonstrated that injection of new heart cells to repair damaged heart tissue resulted in premature death of implanted cells due to the mechanical biaxial squeezing action of the contracting myocardium [38]. An NPR property may be ideally suited for tissue engineering applications that require biaxial expansion; the NPR nature of the scaffold would cause concurrent deformations with the beating of the heart [39]. PEGDA biomaterial was chosen because of its widespread application as scaffolds for tissue engineering of soft tissues. For DMD-PP, prepolymer solution was formed by mixing (PEGDA, Mn = 700) with acrylic acid (AA), quencher, and photoinitiators (Irgacure 2959 and TINUVIN 234 UV-dye). Figure 3.2a highlights the geometry and dimensions of the unit-cell type, the reentrant and cut missing rib designs for NPR behavior, and the intact rib model as PPR control. Rectangular slabs of material were incorporated at the ends of each porous sheet to ensure the mechanical integrity of the sheet for handling during strain testing (Figure 3.2f and g). The arrangement of the ribs relative to one another engender a negative Poisson's ratio by virtue of a combination of rib bending and angular deformations (Figure 3.2c and d). The degree to which each structure deforms depends on unit-cell geometry, the bulk material properties of the ribs, and the direction of loading. The reentrant structure (Figure 3.2a) is formed by changing the four side angles (angle ζ) between the vertices (ribs) in a six-sided honeycomb (hexagon), with some additional modifications [28,40]. Varying angle ζ alters the magnitude of Poisson's ratio,

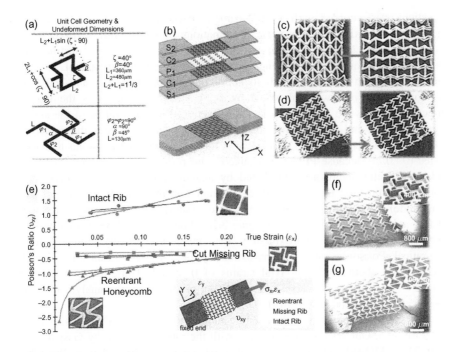

FIGURE 3.2

(a) Unit-cell parameters of reentrant honeycomb and cut missing rib architectures. (b) Schematic of the multilayer scaffolds assembled by stacking single-layer cellular constructs with a connecting layer of vertical posts. (c,d) Side-by-side images show the scaffolds in their (left) undeformed and (right) deformed strain states in response to an applied axial strain. (e) Measured Poisson's ratio as a function of true strain for the single-layer scaffold composed of the reentrant, missing rib, and intact rib (control) unit-cell geometries. Three strain-dependent experiments were performed for each type of single-layer scaffold; each strain test was conducted with a different sample, denoted by color in the plots. (f) SEM images of PEG scaffolds with tunable negative Poisson's ratios. The walls of the unit-cells (denoted as ribs) are approximately 40 μm wide and 100 μm deep.

which gives Poisson's ratio its strain-dependent response. The missing rib model (Figure 3.2a) is formed by removing select ribs from the intact model so the intact form has "missing" ribs [41]. The intact rib meshwork has a positive Poisson's ratio regardless of the direction of loading.

Strain tests were conducted to determine the Poisson's ratios of the single-layer constructs as a function of true (instantaneous) axial strain. The PEG scaffolds were loaded into a homemade strain measurement system. A "pulling" axial tensile stress was applied to the end of the PEG scaffolds using a custom-built strain tester. In-plane movement of the construct in the axial and transverse directions was recorded with a CCD camera. Poisson's ratio was calculated by measuring the axial and transverse deformations of the overall scaffold mesh-works. Figure 3.2e plots the Poisson's ratios of the single-layer constructs for values of true strain ($\varepsilon = \delta L/L$, where ε is true strain, L is length, and δL is incremental change in length) from 0 to approximately 0.2 for each unit-cell type. Poisson's ratio (υ_{xy}) was calculated by $\upsilon_{xy} = -\varepsilon_y/\varepsilon_x$, where x is the loading direction and y is the lateral (transverse) direction. Poisson's ratios of the single-layer reentrant and missing rib constructs were negative, while the intact rib construct (used as a control) was positive for the values of true strain that were tested (0–0.2). Because PEG is not auxetic, our results show that pore geometry induced auxetic behavior as predicted by analytical models. The experimental Poisson's ratios for the single-layer reentrant construct decreased linearly (in magnitude) from approximately −1 to approximately −0.5 for increasing values of true axial strain from 0 to 0.2 (Figure 3.2e). For $L_2/L_1 = 1.33$, which gave an undeformed angle ζ of approximately 40 degrees, the hinging model yields a theoretical Poisson's ratio of approximately −1 at zero strain that linearly decreases to about −0.7 for axial strains of 0 to 0.2. Thus, our experimental values (−1 to −0.5) are very similar to those predicted by the analytical model (−1 to −0.7) for axial strains of 0 to 0.2.

The single-layer missing rib structure demonstrated Poisson's ratios of about −0.3 to −0.5. Our results agree closely with the model reported by Gaspar et al. [42]. The DMD-PP technique can be also extended to multiple layers. We fabricated three-dimensional polymer PEG scaffolds (Figure 3.2b) by stacking two single-layer sheets (C_1 and C_2) with a layer of vertical posts (P_1). The double-layer scaffolds exhibited strain-dependent Poisson's ratios that were very similar to those of the single-layer constructs, which suggested that adding additional layers to a scaffold does not markedly affect the tunability of the Poisson's ratio. In summary, the DMD-PP system was used to construct PEGDA scaffolds, which exhibited tunable in-plane negative Poisson's ratios, consistent with the analytical models found in the literature.

3.3 Hybrid Poisson's ratio biomaterial

DMD-PP was used to design and fabricate scaffolds having adjacent regions of NPR and PPR property using polyethylene glycol (PEG)-based biomaterials. DMD-PP is able to impart precise cellular geometries and deformation mechanisms to tune Poisson's ratio on both single- and dual-layer scaffolds using a side-to-side and a top-to-bottom configuration. The geometry and relevant dimensions of unit cells with negative Poisson's ratio (NPR) and positive Poisson's ratio (PPR) behavior were used to make dynamic photo images for the DMD-PP system (Figure 3.2a). Axial load was applied to both single- and double-layer PEG constructs, and the Poisson's ratio was calculated. With axial loading, the PPR regions on single-layer scaffolds contract, while the NPR regions transversely expand, demonstrating a hybrid NPR/PPR behavior on a single scaffold (Figure 3.3a–c). Addition of multiple layers did not alter the Poisson's ratios' response relative to the hybrid single-layer scaffolds (Figure 3.3d–f). Optical images demonstrate expected deformation behavior; the NPR parts in both the top and bottom layers expand laterally, while the PPR parts shrink with axial loading. Our strain-dependent Poisson's ratio data for the hybrid scaffold match well with the analytical models reported in the literature.

We sought to determine the response of human mesenchymal stem cells (hMSCs) on hybrid scaffolds, since it is an important cell source for regeneration of muscle, cartilage, and fat [43]. PEG-AA scaffolds conjugated with fibronectin and bone marrow–derived hMSCs were seeded on hybrid scaffolds. hMSCs were

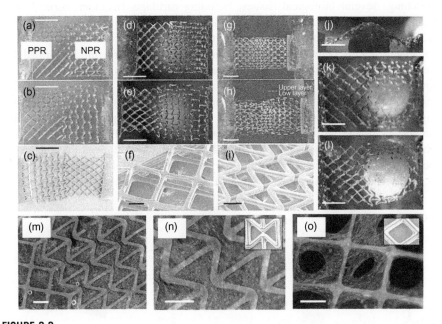

FIGURE 3.3

Optical images of (a,b) single-layer, (d,e) double-layer, and (g,h) switch-layered hybrid PEG scaffolds in response to an applied axial strain. The images show the scaffolds in their (top) undeformed and (bottom) deformed strain states. SEM images of (c) single-layer, (f) double-layer, and (i) switched-layer hybrid scaffolds. Optical images showing NPR behavior on selective regions of top and bottom scaffolds. (j–l) Optical images of out-of-plane loading on hybrid scaffold: A 2-mm ceramic bead was vertically pushed against the hybrid scaffold, and top and side view optical images were captured. The NPR region conforms to the bead surface: Green-actin filaments; Blue-Nuclei; Pink-Scaffold struts. (m–o) Fluorescence microscopy images of hMSC on hybrid scaffolds. Insets show SEM image of unit cell. Scale: (a–e,g,h,j–l) = 1 mm; (f–l,m–o) = 100 μm.

found to readily attach to both the PPR and NPR regions of the hybrid scaffolds, with potential tissue regeneration applications (Figure 3.3m−o). Out-of-plane loading of the hybrid scaffold was assessed using a 2-mm ceramic bead pushing against the hybrid scaffold (Figure 3.3j−l). Loading near the PPR region demonstrates contraction of the PPR region while expansion and conformation of the NPR region to the bead surface. This demonstrates that a biological hybrid patch can be designed for wound management applications especially for treating pressure ulcers.

3.4 Zero Poisson's ratio biomaterials

A ZPR scaffold demonstrates no change in the lateral dimensions, with axial stretching. Several biological tissues, including cartilage, ligament, corneal, and brain, are known to possess Poisson's ratios of nearly zero [44−47]. A variation of the honeycomb model, the semireentrant honeycomb, has been shown to implement a ZPR (Figure 3.4a) [48]. The semireentrant structure is formed by changing the four side angles (angle θ) between the vertices (ribs) in a six-sided honeycomb (hexagon), with some additional modifications [28,40]. Two of the angles have a positive angle, $+\theta$, and the opposite two angles have a negative angle, $-\theta$. Two rib lengths, L_1 and L_2, and angle θ completely constrain the dimensions of the unit-cell. The size of the unit-cell is defined by X_1 and X_2, where $X_1 = 2L_1 \cos\theta$ and $X_2 = 2L_2$. Attard and Grima [48] showed that the semireentrant mesh is isotropic, exhibiting a ZPR for any in-plane normal stresses (σ_x, σ_y) regardless of their directionality. DMD-PP was used to create single- and

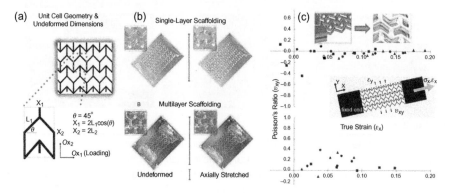

FIGURE 3.4

(a) Unit-cell of semireentrant structure. (b) Optical images showing the deformation of a single- and double-layer zero Poisson's ratio (ZPR) PEG scaffold in response to an axial strain. (c) Plots of Poisson's ratio as a function of true strain for (top) single-layer and (bottom) double-layer scaffolds. Inset shows schematic and SEM image of a two-layer ZPR PEG scaffold.

double-layer scaffolds composed of semireentrant pores whose arrangement and deformation mechanisms contribute to the ZPR (Figure 3.4b). Strain experiments prove ZPR behavior in the single- and double-layered constructs that is accurately predicted by the analytical semireentrant model (Figure 3.4c).

CONCLUSION

DMD-PP is a versatile platform capable of fabricating complex 3D scaffolds with precise microstructures. In this chapter, we focus on the versatility of DMD-PP in fabricating multilayer PEG scaffolds, which exhibit tunable Poisson's ratio behavior, in accordance with existing analytical models found in the literature. Poisson's ratio can be tuned in any photocurable biomaterial without altering the intrinsic elastic modulus property [28]. Tuning Poisson's ratio in a multilayered scaffold, combined with embedded drugs/growth factor, can be extremely versatile for a variety of biomedical applications. Tunable Poisson's ratio scaffolds may be more suitable for emulating the behavior of certain tissues and supporting and transmitting forces to the host site and would likely better integrate with native tissues. Poisson's ratio tuning using DMD-PP is both scale-independent and independent of the choice of structure material for strains in the elastic regime. Lastly, Poisson's ratio behavior can be achieved using a variety of photocurable materials, with potential applications in biomedical applications that require unique strain characteristics.

References

[1]   Blitterswijk CV. Tissue engineering. 1st ed. Boston: Amsterdam; 2008.
[2]   Gibson LJ, Ashby MF. Cellular solids: structure and properties. Cambridge; New York: Cambridge University Press; 1997.
[3]   Khademhosseini A, Langer R. Microengineered hydrogels for tissue engineering. Biomaterials 2007;28(34):5087−92.
[4]   Chen XG, Brodland GW. Mechanical determinants of epithelium thickness in early-stage embryos. J Mech Behav Biomed Mater 2009;2(5):494−501.
[5]   Timmins LH, Wu QF, Yeh AT, Moore JE, Greenwald SE. Structural inhomogeneity and fiber orientation in the inner arterial media. Am J Physiol−Heart C 2010;298 (5):H1537−45.
[6]   Williams JL, Lewis JL. Properties and an anisotropic model of cancellous bone from the proximal tibial epiphysis. J Biomech Eng−T ASME 1982;104(1):50−6.
[7]   Burriesci G, Bergamasco G, inventors; Sorin Biomedica Cardio, assignee. Annuloplasty prosthesis with an auxetic structure. United States, 2007.
[8]   Lin CY, Kikuchi N, Hollister SJ. A novel method for biomaterial scaffold internal architecture design to match bone elastic properties with desired porosity. J Biomech 2004;37(5):623−36.

[9] Lakes R. Materials science: a broader view of membranes. Nature 2001;414 (6863):503−4.

[10] Jackman RJ, Brittain ST, Adams A, Prentiss MG, Whitesides GM. Design and fabrication of topologically complex, three-dimensional microstructures. Science 1998; 280(5372):2089−91.

[11] Veronda DR, Westmann RA. Mechanical characterization of skin—finite deformations. J Biomech 1970;3(1):111−22 [IN119, 123-124].

[12] Baughman RH, Galvao DS. Crystalline networks with unusual predicted mechanical and thermal properties. Nature 1993;365(6448):735−7.

[13] Gardner GB, Venkataraman D, Moore JS, Lee S. Spontaneous assembly of a hinged coordination network. Nature 1995;374(6525):792−5.

[14] Hall LJ, Coluci VR, Galvao DS, Kozlov ME, Zhang M, Dantas SO, et al. Sign change of poisson's ratio for carbon nanotube sheets. Science 2008;320(5875): 504−7.

[15] Lakes R. Foam structures with a negative Poisson's ratio. Science 1987;235(4792): 1038−40.

[16] Choi JB, Lakes RS. Nonlinear properties of metallic cellular materials with a negative Poisson's ratio. J Mater Sci 1992;27(19):5375−81.

[17] Choi JB, Lakes RS. Nonlinear properties of polymer cellular materials with a negative Poisson's ratio. J Mater Sci 1992;27(17):4678−84.

[18] Milton GW. Composite materials with Poisson's ratios close to −1. J Mech Phys Solids 1992;40(5):1105−37.

[19] Alderson KL, Evans KE. The fabrication of microporous polyethylene having a negative Poisson's ratio. Polymer 1992;33(20):4435−8.

[20] Caddock BD, Evans KE. Microporous materials with negative Poisson's ratios. 1. Microstructure and mechanical properties. J Phys D Appl Phys 1989;22(12): 1877−82.

[21] Evans KE, Caddock BD. Microporous materials with negative Poisson's ratios. 2. Mechanisms and interpretation. J Phys D Appl Phys 1989;22(12):1883−7.

[22] Burns S. Negative Poisson's ratio materials. Science 1987;238(4826):551.

[23] Baughman RH, Stafstram S, Cui C, Dantas SO. Materials with negative compressibilities in one or more dimensions. Science 1998;279(5356):1522−4.

[24] Evans KE, Nkansah MA, Hutchinson IJ, Rogers SC. Molecular network design. Nature 1991;353(6340): 124-124.

[25] Rothenburg L, Berlin AI, Bathurst RJ. Microstructure of isotropic materials with negative Poisson's ratio. Nature 1991;354(6353):470−2.

[26] Choi JB, Lakes RS. Nonlinear-analysis of the Poisson's ratio of negative Poisson's ratio foams. J Compos Mater 1995;29(1):113−28.

[27] Evans KE, Alderson A. Auxetic materials: functional materials and structures from lateral thinking. Adv Mater 2000;12(9):617−28.

[28] Gibson LJ, Ashby MF, Schajer GS, Robertson CI. The mechanics of two-dimensional cellular materials. Proc Royal Soc Lond A Mat 1982;382(1782):25−42.

[29] Almgren RF. An isotropic three-dimensional structure with Poisson's ratio = −1. J Elast 1985;15(4):427−30.

[30] Ho M-H, Kuo P-Y, Hsieh H-J, Hsien T-Y, Hou L-T, Lai J-Y, et al. Preparation of porous scaffolds by using freeze-extraction and freeze-gelation methods. Biomaterials 2004;25(1):129−38.

[31] Lannutti J, Reneker D, Ma T, Tomasko D, Farson D. Electrospinning for tissue engineering scaffolds. Materials Science and Engineering: C 2007;27(3):504−9.

[32] Krause B, Diekmann K, van der Vegt NFA, Wessling M. Open nanoporous morphologies from polymeric blends by carbon dioxide foaming. Macromolecules 2002;35 (5):1738−45.

[33] Han LH, Mapili G, Chen S, Roy K. Projection microfabrication of three-dimensional scaffolds for tissue engineering. J Manuf Sci E−T ASME 2008;130(2): pp. 021005-1-4.

[34] Lu Y, Mapili G, Suhali G, Chen SC, Roy K. A digital micro-mirror device-based system for the microfabrication of complex, spatially patterned tissue engineering scaffolds. J Biomed Mater Res A 2006;77A(2):396−405.

[35] Mapili G, Lu Y, Chen SC, Roy K. Laser-layered microfabrication of spatially patterned functionalized tissue-engineering scaffolds. J Biomed Mater Res B 2005;75B (2):414−24.

[36] Soman P, Kelber JA, Lee JW, Wright T, Vecchio KS, Klemke RL, et al. Cancer cell migration within 3D layer-by-layer microfabricated photocrosslinked PEG scaffolds with tunable stiffness. Biomaterials 2012. In Press.

[37] Soman P, Tobe BTD, Lee JW, Winquist A, Singec I, Vecchio KS, et al. Three-dimensional scaffolding to investigate neuronal derivatives of human embryonic stem cells. Biomed Microdevices 2012. In Press.

[38] Teng CJ, Luo J, Chiu RCJ, Shum-Tim D. Massive mechanical loss of microspheres with direct intramyocardial injection in the beating heart: implications for cellular cardiomyoplasty. J Thorac Cardiovasc Surg 2006;132(3):628−32.

[39] Jawad H, Lyon AR, Harding SE, Ali NN, Boccaccini AR. Myocardial tissue engineering. Brit Med Bull 2008;87(1):31−47.

[40] Masters IG, Evans KE. Models for the elastic deformation of honeycombs. Compos Struct 1996;35(4):403−22.

[41] Smith CW, Grima JN, Evans KE. A novel mechanism for generating auxetic behaviour in reticulated foams: missing rib foam model. Acta Mater 2000;48(17): 4349−56.

[42] Gaspar N, Ren XJ, Smith CW, Grima JN, Evans KE. Novel honeycombs with auxetic behaviour. Acta Mater 2005;53(8):2439−45.

[43] Hwang Y, Phadke A, Varghese S. Engineered microenvironments for musculoskeletal differentiation of stem cells. Regen Med 2011;6(4):505−24.

[44] Fatt I. Dynamics of water transport in the corneal stroma. Exp Eye Res 1968;7:402−12.

[45] Jurvelin JS, Arokoski JPA, Hunziker EB, Helminen HJ. Topographical variation of the elastic properties of articular cartilage in the canine knee. J Biomech 2000;33 (6):669−75.

[46] Jurvelin JS, Buschmannf MD, Hunziker EB. Optical and mechanical determination of Poisson's ratio of adult bovine humeral articular cartilage. J Biomechanics 1997;30(3):235−41.

[47] Kyriacou S, Mohamed A, Miller K, Neff S. Brain mechanics for neurosurgery: modeling issues. Biomech Modeling Mechanobiol 2002;1:151−64.

[48] Grima JN, Attard D. Molecular networks with a near zero Poisson's ratio. Phys Status Solidi B 2011;248(1):111−6.

Fabrication of Microscale Hydrogels for Tissue Engineering Applications

Gulden Camci-Unal[1], Pinar Zorlutuna[1], Ali Khademhosseini[1,2]

[1]Center for Biomedical Engineering, Department of Medicine, Brigham and Women's Hospital, Harvard Medical School, Cambridge, MA, 02139, USA. Harvard-MIT Division of Health Sciences and Technology, Massachusetts Institute of Technology, 77 Massachusetts Avenue, Cambridge, MA, 02139, USA; [2]Wyss Institute for Biologically Inspired Engineering, Harvard University, Boston, MA, 02115, USA

CONTENTS

INTRODUCTION

Tissue engineering uses an interdisciplinary approach combining life sciences, medicine, and engineering principles to address the challenges related to replacement and regeneration of tissues and organs [1]. The ultimate goal of tissue engineering is to obtain functional tissues to repair or replace damaged

Biofabrication.

organs. However, a major hurdle in engineering cellular microenvironments is to mimic the structural, mechanical and functional complexity of living tissues. To mimic the complexity that is present in native tissues, cells in artificial constructs should be properly organized [2]. Additionally, cells should be in contact with one another, providing them with adequate signaling [3] and communication [1] to control or direct cellular behavior. Another crucial requirement for the engineered tissues to function properly is vascularization. It is well known that the lack of vascularization may cause issues for diffusion of oxygen and metabolites as well as transportation of waste [1]. To overcome these challenges, various techniques have been developed for the generation of controlled microarchitectures.

Tissue engineered constructs are commonly fabricated by encapsulating cells within hydrogel-based materials. Hydrogels are made of synthetic or natural polymer precursors, which can be crosslinked with suitable reagents [4], resulting in 3D insoluble matrices [5]. Crosslinking of hydogels can be induced with the use of a variety of stimuli, such as chelating ions, pH, temperature, or light exposure. Hydrogels are soft and permeable, and they can absorb high volumes of water [4,6−8]. Hydrogels have the ability to retain water in between their crosslinked chains, which provide them with the flexibility and mechanical properties that are present in the native elastic ECM frameworks [5]. In addition, hydrogels possess porous structures that allow for passage of nutrients, oxygen, metabolites, and waste for encapsulated cells [7,9].

Hydrogels are used in a wide variety of applications because of the ability to tune their swelling, mechanical properties, chemical and physical structures, crosslinking density, diffusivity, and porosity. Due to these reasons, hydrogels are frequently utilized in tissue engineering applications, such as for cell encapsulation [5], delivery of molecules [5], as bioadhesives, and biosensors [4]. Hydrogel scaffolds can also be used to direct cell behavior and promote cellular organization [5]. For this reason, patterned hydrogels may be useful in proteomics, biosensing, drug delivery, diagnostics and tissue engineering [7]. Hydrogels can be generated from natural or synthetic polymers. Some of the most common materials to generate hydrogels are poly-2-hydroxyethyl methacrylate (polyHEMA) [10], poly(ethylene glycol) diacrylate (PEGDA) [4,5,9,11], agarose [4,5,12], alginate [2,4,5], chitosan [5], collagen [5,13], dextran [2], chondroitin sulfate [2], fibrin [4,5,13,14], gelatin [2,13], and hyaluronic acid [2,4,5,9,15].

To control the cellular microenvironment, hydrogels of different features can be fabricated. Traditionally, microscale hydrogels are produced by a number of different techniques, such as photolithography, soft lithography, bioprinting, microfluidics, and emulsification. Additionally, cells can be incorporated into hydrogels using these techniques in a reproducible fashion with defined patterns. It is expected that controlling structural architecture will aid in the generation of properly functioning tissue constructs [1]. For this reason, microscale hydrogels could be useful to understand cell−cell and cell−material interactions. Herein we discuss the most common fabrication techniques to produce microscale hydrogels and then outline a number of different applications for tissue engineering.

4.1 Technologies for fabrication of microengineered hydrogels

To fabricate biomimetic tissue engineered structures, cells can be encapsulated within patterned hydrogels. Such size- and shape-controlled cell-laden hydrogel-based systems can lead to fabrication of engineered organs in the future for clinical purposes. Hydrogel fabrication techniques provide control over spatial resolution of the crosslinking [1]. Furthermore, cell—cell and cell—substrate interactions can be studied with the aid of different hydrogel fabrication techniques. In this section, we provide essential basics for microfabrication of hydrogels for tissue engineering applications.

4.1.1 Photolithography

Photolithography uses a mixture of hydrogel precursor and photoinitiator, which undergo a crosslinking reaction when exposed to ultraviolet (UV) light (Figure 4.1). This technique controls UV light exposure on the hydrogel precursor with a photomask, on which desired patterns are preprinted [4,5]. Once the photo-crosslinking is complete, unreacted polymer is washed out. Photolithography allows a resolution in the range of millimeters to micrometers, providing unique opportunities for tissue engineering research [1]. Photolithography is a low-cost, simple, and time-efficient hydrogel fabrication technique providing consistent pattern fidelity [5]. In addition, it can be used in combination with other techniques to fabricate cell-laden hydrogels. The drawbacks of this technique are the requirement of photocurable polymer precursors [1] and the possibility of DNA damage in the cells due to the photoinitiator, which forms free radicals when exposed to UV light [5].

4.1.2 Soft lithography

Soft lithography is another approach that can be used for fabrication of shape and size-controlled hydrogels (see Figure 4.1). Soft lithography is a popular set of techniques to fabricate microstructures by using elastomeric molds [16]. These techniques can be used in a host of applications, such as biosensors, directing cell adhesion, drug screening, cell sorting, and tissue engineering. Soft lithography enables molding or printing patterns of biological molecules with controlled spatial distribution and topography [2]. This technique is widely used to produce shape-controlled microscale hydrogels with or without the addition of cells in the prepolymer mixture. In soft lithography, a patterned elastomeric mold is placed over a liquid prepolymer solution, which is then allowed to polymerize, providing

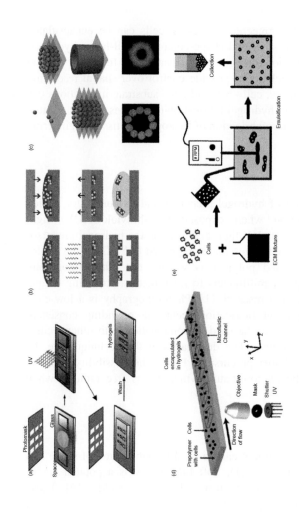

FIGURE 4.1

Fabrication techniques to produce microscale hydrogels. (a) Schematic illustration of the photolithographic approach. Photolithography allows 3D encapsulation of cells within hydrogels by crosslinking the cell-containing prepolymer under UV light. A photomask is used to obtain the desired pattern [88]. (b) The hydrogel fabrication strategy by PDMS molds for soft lithography. Cell-laden hydrogels are prepared by a soft lithographic technique [9]. (c) Formation of a tissue-like construct by a layer-by-layer bioprinting approach [32]. (d) The stop flow lithography (SFL) technique [88]. (e) A typical emulsification experiment to generate hydrogel microparticles [89].

[Reproduced by permission of The Royal Society of Chemistry (a). Reproduced by permission of Elsevier (b). Reproduced by permission of The Royal Society of Chemistry (c,d). Reproduced by permission of John Wiley & Sons, Inc. (e).]

the complementary hydrogel features [5,7]. In this process, liquid fills in the patterns on the elastomeric replica by surface tension and capillary flow [17]. The popularity of this technique is due to a number of advantages, such as its easy accessibility [16,18], simplicity [16,17], cost-effectiveness [2,16,18,19], flexibility [17,19], versatility [19], and time-efficiency [2]. This technique provides precise control over the size and shape of the structures created [16], and it does not require expensive instrumentation [2,20]. Furthermore, soft lithography allows patterning of planar substrates as well as their nonplanar analogs [18]. However, soft lithography has limitations with ionically or chemically crosslinked hydrogel substrates, such as chitosan or alginate [21].

Elastomeric poly(dimethylsiloxane) (PDMS) is widely used to make molds for production of micron-sized features (2−500 μm) [18,22]. PDMS is fabricated by mixing silicon elastomer and curing agent, which is then poured over a micropatterned master and thermally cured. This provides the corresponding replica to use in micropatterning of a variety of polymers to produce hydrogel microstructures [19]. PDMS is a porous biocompatible elastomer, which is successfully used with cell cultures. Furthermore, PDMS is suitable for plasma treatment to change surface hydrophilicity [18]. In addition to PDMS, polyimides, epoxides, and polyurethanes are also used in molding studies. For instance, in one study, polyvinyl alcohol (PVA) was utilized to make microchannels on a silicone mold [23].

4.1.3 Bioprinting

Bioprinting is a computer-aided design (CAD)-based approach, where deposition of cells along with other materials, such as hydrogels, is achieved by a printer-dispensing mechanism in spatially and temporally controlled fashion (see Figure 4.1) [24]. CAD-based technologies have recently become popular in the fabrication of porous scaffolds, especially when cells are seeded into these constructs during the printing stage [25]. 3D complex microenvironments may be generated by utilizing bioprinting systems, where cells are integrated in hydrogels to mimic natural tissues [24]. Therefore, bioprinting could potentially address main challenges related to the generation of spatially controlled 3D complex structures for tissue engineering and regenerative medicine. This approach provides high precision control over the shape and size of the features (printing resolution: tens to hundreds of μm) [26]. It is expected that technological advances and controlled spatial and temporal resolution in the bioprinting processes will make it possible to build organs [24].

The most common bioprinting approaches include inkjet-based cell printing [24,27−32], syringe-based cell deposition [24,33,34], laser writing [24], laser-induced bioprinting [28,35−39], and rapid prototyping [27,40,41] technologies. Inkjet printing is a cost-effective and rapid technique, but it exerts strong mechanical forces on cells while printing [32]. In addition, printing nozzles can get clogged if high cell density is used [28,42]. Such clogging issues in the nozzles could also happen due to ink drying or high polymer viscosity [42]. Laser-assisted bioprinting (LAB), which is another bioprinting technique, is a useful approach to print cells, polymers, or biomolecules. This technique utilizes a laser-absorbing interlayer, which does not let a direct interaction between the ink material and the laser beam occur [42].

In bioprinting-based techniques, cells are accurately placed in predetermined locations [25]. Hyaluronic acid (HA) [24,43], PEG [24], collagen [44], agarose [32], fibrin [45], and alginate [25,27,46] are popular hydrogel precursors to generate tissues by bioprinting for the formation of complex 3D structures. Using such hydrogel precursors with natural origins, bioprinting technlogies are expected to render generation of artificial organs in the future.

4.1.4 Microfluidics

An alternative strategy to fabricate cell-laden hydrogels are microfluidic approaches (see Figure 4.1). Microfluidic systems offer several advantages over conventional techniques for fabrication of microscale hydrogels. Microfluidics enable production of homogeneous droplets [4] with a range of different sizes [1]. However, microfluidic systems are not scalable and allow mainly spherical-shaped particles [21].

Microfluidic systems can be used to obtain *in situ* gelation of prepolymer via electrostatic interactions. For example, Kini et al. [47] demonstrated the fabrication of *in situ* microgels in microfluidic channels. The hydrogels were formed by electrostatic interactions between multivalent citrate anions and cationic polymer poly(allylamine hydrochloride) (PAH) in the microchannels under continuous laminar flow conditions [47]. Flow rate, shear stress, charge ratio, and pH were found to be the main parameters affecting the quality of the microgels formed in the microchannels.

Recently, continuous hydrogel microfibers with tunable physical and chemical properties have been generated by microfluidic platforms [48]. A microfluidic chip was used in combination with a spinning technique to produce spatially controlled multifunctional fibers in micron-scale. These fibers may be created with different topography or chemistry, or contain multiple types of cells enabling various opportunities for clinical applications.

4.1.5 Emulsification

Emulsions can be formed by mixing two immiscible liquid phases to fabricate hydrogel droplets, which could be subsequently crosslinked by photopolymerization or ionic crosslinking [4]. Hydrophilic and hydrophobic phases are commonly used to produce a two-phase system for the emulsification process (see Figure 4.1). To achieve control over the size of hydrogel droplets, the rate of mixing and viscosity of the liquid phase could be varied. Emulsification is a simple technique, and it does not require expensive or complex experimental setups. Although emulsification makes it possible to fabricate different sizes of hydrogel droplets, this technique is limited to mainly spherical shapes, and it gives a somewhat heterogeneous population of hydrogel particles [5].

4.2 **Tissue engineering applications**

Hydrogels have been widely used for tissue engineering research because of their resemblance to native ECM. Since native tissues are comprised of microscale components, hydrogels can be engineered in microscale to generate biomimetic features. These constructs can be used for cell guidance; patterning co-cultures; controlling material characteristics, such as mechanical properties and porosity; and high-throughput fabrication of complex structures.

4.2.1 **Applications of photolithography in tissue engineering**

Photolithography-based techniques are effective for fabricating microscale hydrogels. Cell-laden hydrogels that were shaped using these methods have been investigated for their potential in constructing tissue engineered structures.

Photolithography allows for fabrication of a variety of hydrogels with different sizes, shapes, and physical and chemical properties. Numerous photolabile polymers were explored using this approach. For example, photocrosslinkable dextran was used to fabricate mechanically durable micron-sized hydrogels with different sizes and features under UV light [49]. HEMA, which is a widely known polymer for hydrogel production, was also used to fabricate 3D hydrogel scaffolds by using photolithography [10]. Photolithography is also a useful technique to fabricate hydrogels with a gradient of material properties [50]. For instance, PEGDA was used to induce stiffness gradients by photocrosslinking [51]. The resulting substrates were seeded with macrophages, and cell response against gel rigidity was monitored.

Controlling cytoskeletal and nuclear alignment of cells, which is often studied as cell guidance, is an elegant way of mimicking natural tissues. In one study, cell alignment along hydrogel microchannels was observed in 3D (Figure 4.2) [52]. The alignment took place through remodeling of the methacrylated gelatin hydrogel, and it was hindered when the MMPs of the encapsulated cells did not function. Various other cell types, including fibroblasts (NIH-3T3), human umbilical vein endothelial cells (HUVEC), skeletal muscle cells (C2C12), and cardiac side population cells (CSP), were used in the same study to demonstrate cellular alignment. Similarly, in another study, PEG-based surface-patterned hydrogels were produced by photolithography to control cell guidance [53].

Photolithography has also been used to engineer artificial vasculature. Although planar open hydrogel channels have been made for vascularization purposes, microvascular complexity could be mimicked better by fabricating interconnected channels. To demonstrate that, 3D interconnected microchannels were generated within a multilayered PEG-based hydrogel substrate utilizing a photopatterning approach [54].

Cells in the native tissues are closely packed together regardless of their types. To mimic natural microenvironments, co-cultures of multiple cell types have been used to generate artificial tissue constructs [3]. Traditional approaches for co-cultures were based on mixing different types of cells and seeding them in a random fashion. However, this method is not optimal for engineering multicellular tissues, since the cells in native tissues are highly organized. With the advances in microfabrication technology, researchers are now able to spatially control the localization and organization of cells and have started to investigate their interactions using these platforms as a first step toward functional-engineered tissues. Spatial control over localization of a variety of cell types, including hepatocytes, fibroblasts, chondrocytes, and embryonic liver cells, was achieved through exploiting dielectrophoretic forces exerted on cells within photolabile polymers after subsequent crosslinking [55]. In this study, parallel formation of more than 20,000 size- and shape-controlled cell aggregates was demonstrated within hydrogels, where cells remained viable and were able to differentiate up to two weeks. When the spatial control was used to modulate cell—cell interactions, bovine articular chondrocyte biosynthesis was shown to be affected by the microarchitecture.

Microscale features were also used to modify physical properties of hydrogels in order to render them more compatible for tissue engineering applications. Currently, it is well accepted that the mechanical properties of the surrounding environment of cells are important for their interaction with materials, cellular

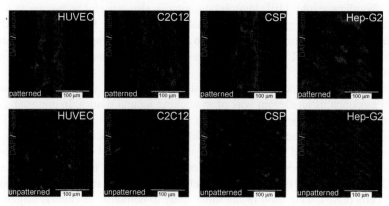

FIGURE 4.2

Cell elongation and alignment in photopatterned cell-laden 5% (w/v) GelMA hydrogels after 5 days of culture. Patterned cells were organized, whereas unpatterned cells grew randomly. DAPI/F-actin staining for patterned and unpatterned HUVEC, C2C12, CSP, and Hep-G2 laden hydrogels are given [52].

Reproduced by permission of Elsevier.

growth, function, and phenotype commitment. To control 3D mechanical properties, microscale rods were fabricated from PEG dimethacrylate (PEGDM) hydrogels using photolithography. As a means of micromechanical cues, these microrods were dispersed in Matrigel to create environments that resemble native tissues. Addition of microrods altered the ECM production, proliferation, and gene expression of the encapsulated fibroblasts [56].

Tissue engineering aims to fabricate implantable 3D functional tissues with spatial control of cells. In one study, Liu et al. [8] used photolithography to pattern hepatocyte (HepG2) encapsulated PEG-based hydrogels with different sizes and shapes with high pattern fidelity. They also generated multilayered hydrogel patterns to mimic the structural complexity of native tissues. In another study, Tsang et al. [57] produced multilayered 3D hepatic tissues with encapsulated cells using a photolithographic approach, characterizing liver-specific functions as the output of measure.

Two photon laser scanning lithography (TP-LSL) was also used to create 3D micropatterned hydrogels to control cellular orientation and migration [58]. PEGDA hydrogels were selectively photocrosslinked to create RGDS patterns in these hydrogels to direct migration of cells in predefined pathways. By this strategy, the concentration and spatial distribution of biomolecules were controlled within hydrogels, and fibroblast migration was guided. Furthermore, this technique enables one to micropattern multiple-cell adhesive ligands in the same hydrogel [59].

In addition to the applications mentioned above, photolithography can be used to fabricate self-assembled hydrogel units loaded with cells. For instance, assemblies of cell-laden hydrogels were formed by micropatterning [60,61]. This approach involves agitation of microgels with desired geometry in a hydrophobic medium, leading to the formation of assembled microgels. This was caused by the hydrophobic forces in the oil-water interface by minimizing free energy at the surface that was exposed to oil. Similarly, in another study, cuboidic hydrogels were self-assembled to form defined shapes [62]. Although highly reproducible, this process could be improved by reducing the surface tension at the liquid interface and increasing the hydrophilicity of the hydrogels. In such studies, assembled hydrogels can be stabilized with a second crosslinking step. As a result, multicomponent cell-laden constructs could be generated through the second crosslinking reaction [63].

4.2.2 Applications of soft lithography in tissue engineering

Soft lithography is another efficient technique to fabricate cell-laden microscale hydrogels with various geometries and dimensions. This technique has been widely utilized to confine microgels into complex architectures for tissue engineering applications. For example, Chiellini et al. [6] functionalized HEMA-based hydrogels for improving cell adhesion and proliferation. They created micropatterns with soft lithography using HEMA to study the effects of surface topology on cellular behavior. In a different study, complex patterns of polyHEMA and PEGDM hydrogel pillars were fabricated by soft lithography [64].

Cell shape, adhesion, and intercellular contact can be controlled by engineering microenvironments, which may aid in regulating cell behavior. For example, in one study, cells are patterned in defined regions of the substrate that allowed cell—cell contact to a certain extent, providing control over cell—biomaterial interactions [19]. In addition to cells, proteins, self-assembled monolayers (SAMs), or other biological molecules can be printed on surfaces by soft lithography [18]. For example, Benedetto et al. [7] produced poly(acrylamide) (PAA) micropatterns using an elastomeric PDMS stamp, producing a platform to deliver proteins on solid supports by stamping. Once the polymerization is complete, the PDMS replica was removed, and hydrogel microgrooves were formed at room temperature with high pattern fidelity, with pattern sizes ranging between 2 and 100 μm.

It is widely known that the spatial position of cells can affect their physiological function [23]. In addition, controlling the shape and geometry of cell-laden hydrogels is critical in the fabrication of tissue-like structures [18] and mimicking native microenvironments [23]. For this reason, micromolding techniques are popular because they allow precise control over the spatial distribution of cells. In one study, NIH-3T3 fibroblasts were encapsulated in photocrosslinkable PEGDA and methacrylated hyaluronic acid (MeHA) precursors, which was placed between a glass slide and a PDMS mold with micropatterns. Following UV exposure, size- and shape-controlled microgels were generated and tested for their biocompatibility with PEGDA and MeHA polymers (Figure 4.3). Hydrated gels were retrieved

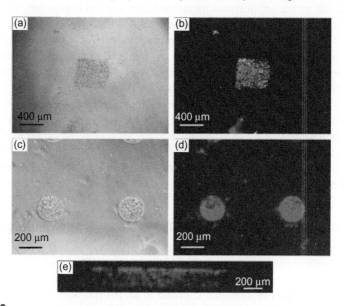

FIGURE 4.3

Phase-contrast and fluorescent images for NIH-3T3 laden MeHA and PEGDA hydrogels stained with a viability assay [9].

Reproduced by permission of Elsevier.

from the PDMS mold and processed for further analysis. In this study, size- and shape-controlled PEGDA and MeHA hydrogels were made with varying cell densities. This approach could be useful to study cell—cell interactions and to generate tissue-like microstructures [9]. The interpretation of signals between cells and the microenvironment provide clues to alter cell behavior in a specific direction. For instance, soft lithography has been used to fabricate engineered microenvironments for studying cell—biomaterial interactions and their effects on cell adhesion in controlled geometries [19,20,65]. Additionally, the effect of pattern size could be critical in cell adhesion affecting cell response and function. To demonstrate the effect of pattern size on cell morphology, Lee et al. [16] used retinal pigment epithelial cells on human lens capsule that was modified by soft lithographic microcontact printing to address the problems related to the loss of cell morphology and function [16]. In an effort to control vasculogenesis with micropatterned hydrogels, PDMS micromolds were filled with HUVEC-laden collagen gels, which resulted in a 3D hollow microvascular tube with control over their orientation and branching when stimulated with VEGF and β-FGF [66]. The dimensions of the microchannels were used to control the tube size.

Micromolds with various geometries and dimensions have been used to confine cells in the hydrogels with different microarchitectures [67—71]. Micromolding for nonadhesive polymers such as polyacrylamide, PEG, and HA was performed to create microscale wells with different geometries (mainly spheroid or toroid) and dimensions. The nonadhesive nature of these wells allowed cells to interact with each other only and self-assemble in the form of aggregates, which can be used to investigate 3D tissue formation. The aggregate size and shape were controlled with the micropatterns, and the aggregates can be easily retrieved from the wells for further applications. For example, methacrylated HA microwells were used to dock cell aggregates. When seeded in microwells, embryonic stem (ES) cells formed spheroids that can be retrieved using mechanical disruption [68]. Another nonadhesive hydrogel, PEG, was used to create microwells with different sizes using the similar micromold-UV crosslinking approach [72—74]. Microwells of different sizes resulted in ES cell aggregates of different sizes, which were further exploited to characterize the effect of aggregate size on their differentiation. These aggregates were shown to be capable of differentiating into lineages from all three germ layers, thus identified as embryoid bodies (EBs) [75]. Smaller EBs (150 μm diameter) were shown to incline more toward vasculogenic differentiation, while larger EBs (450 μm diameter) were differentiating more toward cardiogenic lineage. When two cell types, fibroblasts and HUVECs, were seeded together in polyacrylamide microwells, they self-segregated within these spheroids, forming multilayered structures with an inner fibroblast core and an outer HUVEC layer [76].

Controlling the porosity and interconnected pore structure of a tissue engineered scaffold is crucial both for nourishing the newly forming tissue and for creating the microvasculature in it. In an effort to add vasculature structures into hydrogels, a micromolding approach was used in combination with particulate leaching [77]. Channels were incorporated in the cell-laden hydrogels through leaching of sucrose crystals to create micropores with different dimensions. Sucrose crystals mixed in an agarose solution that was rapidly cooled resulted in a homogeneous distribution of micropores within the hydrogels. The effect of diffusion on cell viability in agarose gels was characterized as a function of porosity. Viability of human hepatic carcinoma cells in these hydrogels correlated with increased diffusivity and porosity.

In the cases where rapid gelling occurs, soft lithography is a popular method of choice to fabricate cell-laden hydrogels. In one study, Franzesi et al. [21] produced cell encapsulated chitosan and alginate hydrogel units and membranes with relief features. The crosslinking was achieved by release of a pH gelling agent or an ionic gelling agent from the mold to chitosan and alginate solutions, respectively. This could be a useful approach to generate 3D co-culture systems to study cellular interactions or for other tissue engineering applications.

4.2.3 Applications of bioprinting in tissue engineering

In the past few years, numerous studies have investigated bioprinting of microscale hydrogels for their potential use in tissue engineering. For instance, in an attempt to fabricate microgels in a high-throughput and automated manner, direct writing techniques were used. A highly accurate (lateral resolution: 10 μm) 3D bioplotting setup was utilized to create microstructured cell-loaded hydrogel constructs with different 3D patterns [78]. The system allowed the embedded hepatocytes to remain viable and functional, while the channels acted as vasculature, which is a promising approach for liver tissue engineering applications. In another study, a bioprinter was used to generate 3D patches from collagen by encapsulating smooth muscle cells (dimensions: 5 mm × 5 mm × 81 um). The printing system used mechanical valves that enabled printing high-viscosity polymers loaded with cells [79]. With this technique, the drawbacks of commercially available inkjet printers, including clogging and low cell viability, can be addressed. In addition, high-throughput droplet formation was achieved with uniformly distributed cells, allowed for high cell viability for at least 14 days. In another study, a combination of bioplotting with laser-induced crosslinking, referred to as biological laser printing (BioLP), was used to microfabricate endothelial cell—laden hydrogel droplets that contained five to seven cells in each spot (70 μm diameter) [80].

Hydrogel-injecting bioprinting systems are popular for manufacturing cell-laden tissue-like constructs. During the hydrogel fabrication steps, there are a number of important factors that affect the printing quality, such as injection velocity, printer nozzle diameter, or flow rate of the polymer [24]. For example, Song et al. [24] demonstrated the effects of printing parameters on pattern width and optimized these parameters to obtain minimum pattern width for tissue engineering applications. Another popular bioprinting approach is cell writing. As an example, endothelial cell—laden 3D alginate hydrogel constructs were bioprinted utilizing a cell writing platform [27]. In another report, scaffolds from porous alginate hydrogels were produced, including

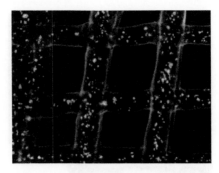

FIGURE 4.4

Optical and fluorescent images of a bioprinted scaffold after 14 culture [25].

Reproduced by permission of ASME.

endothelial cells in the printing polymer mixture (Figure 4.4) [25]. Similarly, Khalil et al. fabricated polymer-based scaffolds with complex 3D architectures by a direct cell writing approach [81]. In a different study, neural cell—laden collagen gels were generated [44] using a bioprinting setup [82]. A similar approach was demonstrated earlier, where rat embryonic neurons and astrocytes were printed in a layer-by-layer fashion to fabricate 3D collagen-based scaffolds [44].

Natural hydrogel precursors are widely incorporated in bioprinting processes. For instance, the organizations of HUVEC and rabbit carcinoma cells in 3D alginate hydrogels were achieved by utilizing a laser-assisted bioprinting technique [83]. In addition, Matrigel and fibrin were used as the gel phase. The resolution of printed cells was determined by manipulating the printing parameters, such as laser printing speed, viscosity of the bioink, or laser energy [83]. Furthermore, HA- and gelatin-based hydrogels were used to encapsulate HepG2 C3A, NIH-3T3, and Int-407 cells using a bioprinting approach [43]. Additionally, a cellularized tubular construct was fabricated by the same printing system.

4.2.4 Applications of microfluidics in tissue engineering

Microfluidic approaches were developed to generate size- and shape-controlled hydrogel droplets under continuous flow conditions. In addition to fabrication of plain hydrogels, cells can be incorporated into polymers utilizing microfluidic systems. For instance, 3D encapsulation of cells within microscale hydrogels has previously been carried out by using microfluidic techniques under continuous flow. In one study, crosslinking of alginate was performed in a microfluidic device to yield a range of droplet sizes (60–100 µm) [84]. Sodium alginate and calcium carbonate ($CaCO_3$) particles were introduced in the aqueous phase in the central channel and mixed with the continuous oil phase, which contained the acetic acid in the side channels. This technique was intended to be used for uniform encapsulation of cells or other biological molecules. Similarly, microscale

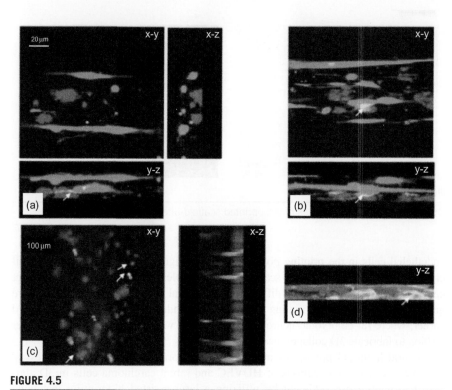

FIGURE 4.5

Fluorescent images of two- or three-layer structures obtained by a layer-by-layer microfluidic setup (a and b) and random mixture (d). Different colors (red, green, or blue) were used to stain cells before patterning with matrices [86].

Reproduced by permission of Elsevier.

cell–laden hydrogel beads were fabricated from alginate and $CaCO_3$ nanoparticles in a microfluidic system [85]. Formation of 90 to 150 μm hydrogel droplets was achieved by the internal gelation principle in micron-sized PDMS channels. A syringe pump was used to produce the hydrogel droplets at a T-junction point in the PDMS device. To fabricate the hydrogel droplets, an emulsion was produced by mixing a corn oil and sodium alginate solution that contained $CaCO_3$ nanoparticles. Release of $CaCO_3$ enabled the crosslinking of alginate hydrogel beads, providing a narrow and consistent distribution of particle size. In this study, flow parameters were used to control the shape and size of the hydrogels. Jurkat cells were successfully encapsulated in these alginate hydrogel beads and cultured, which yielded in high cell viability.

Living tissues and organs are composed of complex architectures, and cells are embedded within layers of tissues. For this reason, engineered environments with different cell types may mimic the native tissue function. For example, layer-by-layer 3D hierarchical tissue constructs were generated by a microfluidic approach (Figure 4.5) [86]. In this study, human lung fibroblasts,

HUVECs, and human umbilical vein smooth muscle cells were encapsulated in Matrigel, fibrin, collagen, or collagen-chitosan polymers, forming multilayered 3D tissue-like structures inside microchannels for potential arterial tissue engineering applications.

In addition to ionic crosslinking, photocurable hydrogel precursors have been used to produce cell encapsulated microstructures in microfluidic systems. In one study, 3D fibroblast-laden PEGDA microgels of different thicknesses were patterned in a microfluidic channel by Cheung et al. [87], which could potentially be useful to study cell behavior. Stop-flow lithography (SFL) is another microfluidic technique to fabricate cell encapsulated microscale hydrogels under continuous flow conditions (see Figure 4.1) [88]. In the SFL setup, UV light is located at the bottom of the microfluidic system to induce crosslinking of photocurable hydrogel precursors through a photomask. To produce size- and shape-controlled cell-laden hydrogel droplets, cells were first resuspended in the prepolymer, and this mixture was allowed to flow through the microfluidic channels. When the prepolymer was exposed to the UV light through a photomask, the resulting microgels were collected from the outlet reservoir of the device. These cell-laden hydrogels could be used to generate 3D structures to mimic native tissues.

4.2.5 Applications of emulsification in tissue engineering

In addition to the techniques mentioned in previous sections, emulsification can be used to fabricate cell-laden micron-sized hydrogels for a number of tissue engineering applications. For instance, a collagen type I and agarose mixture was used to encapsulate human mesenchymal stem cells (hMSC) by an emulsification approach (Figure 4.6) [89]. A warm cell–hydrogel mixture was poured into a liquid PDMS phase and stirred with a mixer to form the cell-encapsulated beads. Subsequently, an ice bath was placed around the mixing vessel to induce gelation of agarose. hMSC beads were then recovered by centrifugation and cultured to monitor osteoblastic differentiation.

In addition to encapsulation within hydrogels, cells have been seeded on hydrogel beads formed by emulsification techniques. For example, human fetal osteoblasts (hFOBs) and human dermal fibroblasts (HDFs) were seeded on gellan-based hydrogels beads (300−850 μm), which were formed in a water-in-oil emulsion setup [90]. To render the gellan microcarriers cell adhesive, a gelatin coating was applied before the cell seeding step. This system allowed adhesion, proliferation, and differentiation of cells with potential application in musculoskeletal regenerative medicine research. Similarly, injectable cell-laden hydrogel microcarriers were produced by emulsification with the potential use in bone tissue engineering [91]. In this study, gelatin-coated gellan microspheres were formed in a water-in-oil emulsion system. After a one-day postseeding of hMSCs or human fetal osteoblasts (hFOBs) on the microcarriers, they were encapsulated in agarose hydrogel and cultured in the corresponding media to study survival and osteogenic differentiation. Another example for stem cell differentiation was

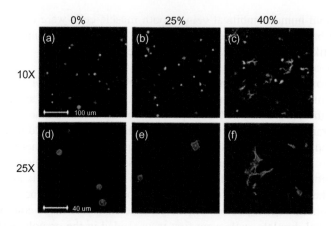

FIGURE 4.6

Fluorescent images demonstrating the morphology of hMSCs after 8 days in culture in alginate beads with varying collagen concentrations generated by an emulsification approach [89].

Reproduced by permission of John Wiley & Sons, Inc.

studied by Dang et al. [92]. In this report, microcapsules of human ES cell aggregates were generated by an emulsification process. Cell-laden agarose was mixed with dimethylpolysiloxane, and droplets were fabricated by a droplet maker device. Subsequently, cell growth and differentiation were studied.

The emulsification approach has also been utilized in other cell differentiation studies, such as in cardiomyogenesis. In one example, aggregated ESCs were formed and encapsulated in ionically crosslinked alginate capsules [93]. The mass production of EBs in alginate hydrogel beads was obtained, and their survival and differentiation into cardiomyocytes were monitored.

CONCLUSION

There are a vast number of techniques available to produce cell-laden hydrogels to manipulate cellular microenvironments. These approaches are crucial to the study of cell–cell or cell–biomaterial interactions, as well as cell–soluble factor interactions. Microscale hydrogel fabrication techniques enable production of tissue-like constructs, which could potentially lead to the formation of artificial organs. These techniques are anticipated to shed light on cell–cell and cell–biomaterial interactions, providing better understanding at the molecular level. To produce 3D tissue constructs with complexities similar to living tissues, the fabrication techniques should be optimized in such a way that they allow for the formation of microarchitectures, which can be used to fabricate vascular structures and properly assembled cells. Accordingly, these constructs are expected to be both structural and functional with potential *in vivo* applications in tissue engineering research. Altogether, fabrication techniques to produce cell-encapsulated microscale hydrogels will improve the ability to mimic native environments. Such developments create significant impacts in tissue engineering and regenerative medicine, allowing for the production of artificial tissue constructs with improved mechanical and structural features [2]. It is envisioned that microfabrication techniques will address major challenges in medicine, life sciences, and tissue engineering areas in the future.

Acknowledgments

This work was supported by the National Science Foundation (DMR0847287), the Institute for Soldier Nanotechnologies, the U.S. Army Corps of Engineers, and the National Institutes of Health (AR057837; DE019024; HL092836; HL099073).

References

[1] Khademhosseini A, Langer R. Microengineered hydrogels for tissue engineering. Biomaterials 2007;28:5087–92.

[2] Khademhosseini A, Du Y, Rajalingam B, Vacanti JP, Langer R. Microscale technologies for tissue engineering In: Polak J, editor. Advances in tissue engineering. 1st ed. London: Imperial College Press; 2007.

[3] Kaji H, Camci-Unal G, Langer R, Khademhosseini A. Engineering systems for the generation of patterned co-cultures for controlling cell-cell interactions. Biochim Biophys Acta 2011;1810:239–50.

[4] Rivest C, Morrison DWG, Ni B, Rubin J, Yadav V, et al. Microscale hydrogels for medicine and biology: synthesis, characteristics and applications. J Mech Mater Struct 2007;2:1103–19.

[5] Slaughter BV, Khurshid SS, Fisher OZ, Khademhosseini A, Peppas NA. Hydrogels in regenerative medicine. Adv Mater 2009;21:3307–29.

[6] Chiellini F, Bizzarri R, Ober CK, Schmaljiohann D, Yu T, et al. Surface patterning and biological evaluation of semi-interpenetrated poly(HEMA)/poly(alkyl β-malolactonate)s. Macromol Symp 2003;197:369–80.

[7] Di Benedetto F, Biasco A, Pisignano D, Cingolani R. Patterning polyacrylamide hydrogels by soft lithography. Nanotechnology 2005;16:S165–70.

[8] Liu VA, Bhatia SN. Three-dimensional photopatterning of hydrogels containing living cells. Biomed Microdevices 2002;4:257–66.

[9] Yeh J, Ling Y, Karp JM, Gantz J, Chandawarkar A, et al. Micromolding of shape-controlled, harvestable cell-laden hydrogels. Biomaterials 2006;27:5391–8.

[10] Bryant SJ, Cuy JL, Hauch KD, Ratner BD. Photo-patterning of porous hydrogels for tissue engineering. Biomaterials 2007;28:2978–86.

[11] Ivanov I, Schaab C, Planitzer S, Teichmann U, Machl A, et al. DNA microarray technology and antimicrobial drug discovery. Pharmacogenomics 2000;1: 169–78.

[12] Khademhosseini A, May MH, Sefton MV. Conformal coating of mammalian cells immobilized onto magnetically driven beads. Tissue Eng 2005;11:1797–806.

[13] Malafaya PB, Silva GA, Reis RL. Natural-origin polymers as carriers and scaffolds for biomolecules and cell delivery in tissue engineering applications. Adv Drug Deliv Rev 2007;59:207–33.

[14] Sakiyama SE, Schense JC, Hubbell JA. Incorporation of heparin-binding peptides into fibrin gels enhances neurite extension: an example of designer matrices in tissue engineering. Faseb J 1999;13:2214–24.

[15] Burdick JA, Chung C, Jia XQ, Randolph MA, Langer R. Controlled degradation and mechanical behavior of photopolymerized hyaluronic acid networks. Biomacromolecules 2005;6:386–91.

[16] Lee CJ, Blumenkranz MS, Fishman HA, Bent SF. Controlling cell adhesion on human tissue by soft lithography. Langmuir 2004;20:4155–61.

[17] Kim P, Du Y, Khademhosseini A, Langer R, Suh KY. Unconventional patterning methods forbionems. In: Rogers, JA, Lee, HH, editors. Unconventional nanopatterning techniques and applications. 1st ed. Hoboken, New Jersey: John Wiley & Sons, Inc.; 2008.

[18] Kane RS, Takayama S, Ostuni E, Ingber DE, Whitesides GM. Patterning proteins and cells using soft lithography. Biomaterials 1999;20:2363–76.

[19] Shin H. Fabrication methods of an engineered microenvironment for analysis of cell-biomaterial interactions. Biomaterials 2007;28:126–33.

[20] Whitesides GM, Ostuni E, Takayama S, Jiang X, Ingber DE. Soft lithography in biology and biochemistry. Annu Rev Biomed Eng 2001;3:335–73.

[21] Franzesi GT, Ni B, Ling YB, Khademhosseini A. A controlled-release strategy for the generation of cross-linked hydrogel microstructures. J Am Chem Soc 2006;128:15064–5.

[22] Xia YN, Whitesides GM. Soft lithography. Annu Rev Mater Sci 1998;28:153–84.

[23] Cheng C-M, LeDuc PR. Micropatterning polyvinyl alcohol as a biomimetic material through soft lithography with cell culture. Mol Biosyst 2006;2:299–304.

[24] Song SJ, Choi J, Park YD, Lee JJ, Hong SY, et al. A three-dimensional bioprinting system for use with a hydrogel-based biomaterial and printing parameter characterization. Artif Organs 2010;34:1044–8.

[25] Khalil S, Sun W. Bioprinting endothelial cells with alginate for 3D tissue constructs. J Biomech Eng 2009;131:111002.

[26] Ovsianikov A, Gruene M, Pflaum M, Koch L, Maiorana F, et al. Laser printing of cells into 3D scaffolds. Biofabrication 2010;2:0141041–7.

[27] Buyukhatipoglu K, Chang R, Sun W, Clyne AM. Bioprinted nanoparticles for tissue engineering applications. Tissue Eng Part C-Methods 2010;16:631–42.

[28] Guillotin B, Guillemot F. Cell patterning technologies for organotypic tissue fabrication. Trends Biotechnol 2011;29:183–90.

[29] Varghese D, Deshpande M, Xu T, Kesari P, Ohri S, et al. Advances in tissue engineering: cell printing. J Thorac Cardiovasc Surg 2005;129:470–2.

[30] Boland T, Mironov V, Gutowska A, Roth EA, Markwald RR. Cell and organ printing 2: fusion of cell aggregates in three-dimensional gels. Anat Rec Part A—Discov Mol Cell Evol Biol 2003;272A:497–502.

[31] Cui XF, Boland T. Human microvasculature fabrication using thermal inkjet printing technology. Biomaterials 2009;30:6221–7.

[32] Mironov V, Prestwich G, Forgacs G. Bioprinting living structures. J Mater Chem 2007;17:2054−60.

[33] Yan YN, Xiong Z, Hu YY, Wang SG, Zhang RJ, et al. Layered manufacturing of tissue engineering scaffolds via multi-nozzle deposition. Mater Lett 2003;57:2623−8.

[34] Ang TH, Sultana FSA, Hutmacher DW, Wong YS, Fuh JYH, et al. Fabrication of 3D chitosan-hydroxyapatite scaffolds using a robotic dispensing system. Mater Sci Eng C-Biomim Supramol Syst 2002;20:35−42.

[35] Gruene M, Deiwick A, Koch L, Schlie S, Unger C, et al. Laser printing of stem cells for biofabrication of scaffold-free autologous grafts. Tissue Eng Part C—Methods 2010;17:79−87.

[36] Othon CM, Wu XJ, Anders JJ, Ringeisen BR. Single-cell printing to form three-dimensional lines of olfactory ensheathing cells. Biomed Mater 2008;3:034101.

[37] Catros S, Guillotin B, Bacakova M, Fricain JC, Guillemot F. Effect of laser energy, substrate film thickness and bioink viscosity on viability of endothelial cells printed by laser-assisted bioprinting. Appl Surf Sci 2010;257:5142−7.

[38] Wu PK, Ringeisen BR. Development of human umbilical vein endothelial cell (HUVEC) and human umbilical vein smooth muscle cell (HUVSMC) branch/stem structures on hydrogel layers via biological laser printing (BioLP). Biofabrication 2010;2: 014111.

[39] Ringeisen BR, Kim H, Barron JA, Krizman DB, Chrisey DB, et al. Laser printing of pluripotent embryonal carcinoma cells. Tissue Eng 2004;10:483−91.

[40] Cohen DL, Malone E, Lipson H, Bonassar LJ. Direct freeform fabrication of seeded hydrogels in arbitrary geometries. Tissue Eng 2006;12:1325−35.

[41] Wang XH, Yan YN, Pan YQ, Xiong Z, Liu HX, et al. Generation of three-dimensional hepatocyte/gelatin structures with rapid prototyping system. Tissue Eng 2006;12:83−90.

[42] Guillemot F, Souquet A, Catros S, Guillotin B. Laser-assisted cell printing: principle, physical parameters versus cell fate and perspectives in tissue engineering. Nanomedicine 2010;5:507−15.

[43] Skardal A, Zhang JX, McCoard L, Xu XY, Oottamasathien S, et al. Photocrosslinkable hyaluronan-gelatin hydrogels for two-step bioprinting. Tissue Eng Part A 2010;16:2675−85.

[44] Lee W, Pinckney J, Lee V, Lee JH, Fischer K, et al. Three-dimensional bioprinting of rat embryonic neural cells. Neuroreport 2009;20:798−803.

[45] Guillotin B, Souquet A, Catros S, Duocastella M, Pippenger B, et al. Laser assisted bioprinting of engineered tissue with high cell density and microscale organization. Biomaterials 2010;31:7250−6.

[46] Buyukhatipoglu K, Chang R, Sun W, Clyne AM. Bioprinted nanoparticles for tissue engineering. Tissue Eng Part C-Methods 2010;16:631−42.

[47] Kini GC, Lai J, Wong MS, Biswal SL. Microfluidic formation of ionically cross-linked polyamine gels. Langmuir 2010;26:6650−6.

[48] Kang E, Jeong GS, Choi YY, Lee KH, Khademhosseini A, et al. Digitally tunable physicochemical coding of material composition and topography in continuous microfibres. Nature Mater 2011;10:877−83.

[49] Lo CW, Jiang HR. Photopatterning and degradation study of dextran-glycidyl methacrylate hydrogels. Polym Eng Sci 2010;50:232−9.

[50] Burdick JA, Khademhosseini A, Langer R. Fabrication of gradient hydrogels using a microfluidics/photopolymerization process. Langmuir 2004;20:5153−6.

[51] Nemir S, Hayenga HN, West JL. PEGDA hydrogels with patterned elasticity: novel tools for the study of cell response to substrate rigidity. Biotechnol Bioeng 2010;105:636−44.

[52] Aubin H, Nichol JW, Hutson CB, Bae H, Sieminski AL, et al. Directed 3D cell alignment and elongation in microengineered hydrogels. Biomaterials 2010;31:6941−51.

[53] Hahn MS, Taite LJ, Moon JJ, Rowland MC, Ruffino KA, et al. Photolithographic patterning of polyethylene glycol hydrogels. Biomaterials 2006;27:2519−24.

[54] Chiu YC, Larson JC, Perez-Luna VH, Brey EA. Formation of microchannels in poly (ethylene glycol) hydrogels by selective degradation of patterned microstructures. Chem Mater 2009;21:1677−82.

[55] Albrecht DR, Underhill GH, Wassermann TB, Sah RL, Bhatia SN. Probing the role of multicellular organization in three-dimensional microenvironments. Nature Methods 2006;3:369−75.

[56] Ayala P, Lopez JI, Desai TA. Microtopographical cues in 3D attenuate fibrotic phenotype and extracellular matrix deposition: implications for tissue regeneration. Tissue Eng Part A 2010;16:2519−27.

[57] Tsang VL, Chen AA, Cho LM, Jadin KD, Sah RL, et al. Fabrication of 3D hepatic tissues by additive photopatterning of cellular hydrogels. FASEB J 2007;21: 790−801.

[58] Lee SH, Moon JJ, West JL. Three-dimensional micropatterning of bioactive hydrogels via two-photon laser scanning photolithography for guided 3D cell migration. Biomaterials 2008;29:2962−8.

[59] Hoffmann JC, West JL. Three-dimensional photolithographic patterning of multiple bioactive ligands in poly(ethylene glycol) hydrogels. Soft Matter 2010;6: 5056−63.

[60] Du Y, Lo E, Vidula MK, Khabiry M, Khademhosseini A. Method of bottom-up directed assembly of cell-laden microgels. Cell Mol Bioeng 2008;1:157−62.

[61] Zamanian B, Masaeli M, Nichol JW, Khabiry M, Hancock MJ, et al. Interface directed self assembly of cell-laden microgels. Small 2010;6:937−44.

[62] Du Y, Ghodousi M, Lo E, Vidula MK, Emiroglu O, et al. Surface-directed assembly of cell-laden microgels. Biotechnol Bioeng 2010;105:655−62.

[63] Du Y, Lo E, Ali S, Khademhosseini A. Directed assembly of cell-laden microgels for fabrication of 3D tissue constructs. Proc Natl Acad Sci USA 2008;105:9522−7.

[64] Chandra D, Taylor JA, Yang S. Replica molding of high-aspect-ratio (sub-)micron hydrogel pillar arrays and their stability in air and solvents. Soft Matter 2008;4:979−84.

[65] Geissler M, Xia Y. Patterning: principles and some new developments. Adv Mater 2004;16:1249−69.

[66] Raghavan S, Nelson CM, Baranski JD, Lim E, Chen CS. Geometrically controlled endothelial tubulogenesis in micropatterned gels. Tissue Eng Part A 2010;16:2255−63.

[67] Nichol JW, Koshy ST, Bae H, Hwang CM, Yamanlar S, et al. Cell-laden microengineered gelatin methacrylate hydrogels. Biomaterials 2010;31:5536−44.

[68] Khademhosseini A, Eng G, Yeh J, Fukuda J, Blumling 3rd J, et al. Micromolding of photocrosslinkable hyaluronic acid for cell encapsulation and entrapment. J Biomed Mater Res A 2006;79:522−32.

[69] Suh KY, Khademhosseini A, Yang JM, Eng G, Langer R. Soft lithographic patterning of hyaluronic acid on hydrophilic substrates using molding and printing. Adv Mater 2004;16:584−8.

[70] Revzin A, Russell RJ, Yadavalli VK, Koh WG, Deister C, et al. Fabrication of poly (ethylene glycol) hydrogel microstructures using photolithography. Langmuir 2001;17:5440−7.

[71] Albrecht DR, Tsang VL, Sah RL, Bhatia SN. Photo- and electropatterning of hydrogel-encapsulated living cell arrays. Lab Chip 2005;5:111−8.

[72] Karp JM, Yeh J, Eng G, Fukuda J, Blumling J, et al. Controlling size, shape and homogeneity of embryoid bodies using poly(ethylene glycol) microwells. Lab Chip 2007;7:786−94.

[73] Moeller HC, Mian MK, Shrivastava S, Chung BG, Khademhosseini A. A microwell array system for stem cell culture. Biomaterials 2008;29:752−63.

[74] Choi YY, Chung BG, Lee DH, Khademhosseini A, Kim JH, et al. Controlled-size embryoid body formation in concave microwell arrays. Biomaterials 2010;31:4296−303.

[75] Hwang YS, Chung BG, Ortmann D, Hattori N, Moeller HC, et al. Microwell-mediated control of embryoid body size regulates embryonic stem cell fate via differential expression of WNT5a and WNT11. Proc Natl Acad Sci USA 2009;106:16978−83.

[76] Napolitano AP, Chai P, Dean DM, Morgan JR. Dynamics of the self-assembly of complex cellular aggregates on micromolded nonadhesive hydrogels. Tissue Eng 2007;13:2087−94.

[77] Park JH, Chung BG, Lee WG, Kim J, Brigham MD, et al. Microporous cell-laden hydrogels for engineered tissue constructs. Biotechnol Bioeng 2010;106:138−48.

[78] Yan YN, Wang XH, Pan YQ, Liu HX, Cheng J, et al. Fabrication of viable tissue-engineered constructs with 3D cell-assembly technique. Biomaterials 2005;26:5864−71.

[79] Moon S, Hasan SK, Song YS, Xu F, Keles HO, et al. Layer by layer three-dimensional tissue epitaxy by cell-laden hydrogel droplets. Tissue Eng Part C—Methods 2010;16:157−66.

[80] Guillemot F, Souquet A, Catros S, Guillotin B, Lopez J, et al. High-throughput laser printing of cells and biomaterials for tissue engineering. Acta Biomater 2010;6:2494−500.

[81] Khalil S, Nam J, Sun W. Multi-nozzle deposition for construction of 3D biopolymer tissue scaffolds. Rapid Prototyping J 2005;11:9−17.

[82] Lee W, Debasitis JC, Lee VK, Lee JH, Fischer K, et al. Multi-layered culture of human skin fibroblasts and keratinocytes through three-dimensional freeform fabrication. Biomaterials 2009;30:1587−95.

[83] Mironov V, Reis N, Derby B. Bioprinting: a beginning. Tissue Eng 2006;12:631−4.

[84] Zhang H, Tumarkin E, Sullan RMA, Walker GC, Kumacheva E. Exploring micro-fluidic routes to microgels of biological polymers. Macromol Rapid Commun 2007;28:527−38.

[85] Tan WH, Takeuchi S. Monodisperse alginate hydrogel microbeads for cell encapsulation. Adv Mater 2007;19:2696−701.

[86] Tan W, Desai TA. Layer-by-layer microfluidics for biomimetic three-dimensional structures. Biomaterials 2004;25:1355−64.

[87] Cheung YK, Gillette BM, Zhong M, Ramcharan S, Sia SK. Direct patterning of composite biocompatible microstructures using microfluidics. Lab Chip 2007; 7:574−9.

[88] Panda P, Ali S, Lo E, Chung BG, Hatton TA, et al. Stop-flow lithography to generate cell-laden microgel particles. Lab Chip 2008;8:1056−61.

[89] Batorsky A, Liao JH, Lund AW, Plopper GE, Stegemann JP. Encapsulation of adult human mesenchymal stem cells within collagen-agarose microenvironments. Biotechnol Bioeng 2005;92:492−500.

[90] Wang CM, Gong YH, Lin YM, Shen JB, Wang DA. A novel gellan gel-based microcarrier for anchorage-dependent cell delivery. Acta Biomaterialia 2008;4:1226−34.

[91] Wang CM, Gong YH, Zhong Y, Yao YC, Su K, et al. The control of anchorage-dependent cell behavior within a hydrogel/microcarrier system in an osteogenic model. Biomaterials 2009;30:2259−69.

[92] Dang SM, Gerecht-Nir S, Chen J, Itskovitz-Eldor J, Zandstra PW. Controlled, scalable embryonic stem cell differentiation culture. Stem Cells 2004;22:275−82.

[93] Magyar JP, Nemir M, Ehler E, Suter N, Perriard JC, et al. Mass production of embryoid bodies in microbeads. Ann N Y Acad Sci 2001;944:135−43.

Polymeric Membranes for the Biofabrication of Tissues and Organs

5

Antonietta Messina[1,2] **and Loredana De Bartolo**[2]

[1]Department of Chemical Engineering and Materials University of Calabria, via P. Bucci cubo 45/A, I-87030 Rende (CS), Italy; [2]Institute on Membrane Technology, National Research Council of Italy, ITM-CNR, c/o University of Calabria, via P. Bucci cubo 17/C, I-87030 Rende (CS), Italy

CONTENTS

INTRODUCTION

In the past three decades, many types of natural and synthetic biodegradable materials have been used for the development of new medical devices that can act as support for promoting the regeneration of tissues or replacing failing or malfunctioning organs. The importance of the morphological and physico-chemical properties of the biomaterials in cell interactions has been demonstrated. Surface free energy, electric charge, and morphology might all affect the cell attachment and behavior, and when correctly developed, they could support cell processes that build up a new functioning tissue [1]. Several natural and synthetic degradable polymers, including collagen, chitosan, hyaluronic acid, polyglycolic acid (PGA), poly(D-L-)lactic-co-glycolic acid (PLGA), poly-L-lactic acid (PLLA), polycaprolactone (PCL), and polyurethane (PU), have been used for many engineering purposes, due to their characteristics of biodegradability and biocompatibility. Nowadays in particular, synthetic polymeric membranes in various configurations (flat, capillary, hollow fiber, etc.) are widely used in innovative biomedical devices, thanks their biocompatibility and selective permeability properties. In the field of bioartificial organs, in which design, modification, growth, and maintenance of living tissue are requested, the polymeric membranes play an important role chemically and mechanically, supporting the cell behavior in terms of adhesion and growth. This is why, despite the considerable progress in the field of biomaterials science, continuous development of new insights is required for the creation of an implantable tissue engineered system based on polymeric membranes with suitable mass transport and surface properties [2,3].

Several studies have been focused on the development of bio-artificial systems based on the use of biomaterials for the in vitro engineering of living tissues: cultured cells are grown on bioactive degradable substrates (scaffolds) that provide the physical and chemical cues to guide their differentiation and assembly into three-dimensional structures. Already, one can infer the major challenge in that field: finding the perfect combination of cells (culture conditions and growth factors are important for the production of extracellular matrix and its long-term preservation) and biomaterials as artificial support for cellular components to reconstruct lost tissue or a tissue function by a regenerative process, avoiding clinical problems like stress-shielding, allergic reactions, wear particles, and chronic inflammatory reactions. With specific physical, mechanical, and biological properties, the scaffolds act as a substrate for cellular growth, proliferation, and support for new tissue formation. Then, independently of the form used for the final bio-hybrid device, they need to imitate in their composition and/or structure the native and physiological conditions for the tissue-specific cells to perform complex biochemical functions, including adaptive control and the replacement of normal living tissues [4,5].

5.1 Properties of polymeric membranes

On many parameters, polymeric materials satisfy the requirements of biomedical applications, thanks to their resemblance with the natural macromolecules. Whether it's natural or synthetic, a polymer is an organic compound formed by recurring building blocks linked together in a specific manner through covalence bonds. These small units, named monomers, matched with a specific industrial processing method, allow the control of the final polymeric structure in terms of chemistry (linear, branched, reticulate chain) and morphology (amorphous or crystalline phase). All these aspects eventually affect the mechanical and physical properties of the materials and their final applications. Molecular weight, polymerization degree, organization of the multiple chains, and then the response of the polymer to physico-chemical energy in the form of temperature and mechanical strength and pH are the parameters that determine the choice of a particular technique or a processing method suitable for obtaining the final form of the biomaterial for a specific biomedical application. But above all these forms, the polymeric membranes are now among the most promising. Polymeric membranes can be defined as selective barriers that separate two different phases, allowing the transport of some species with respect to other molecules. They can be classified according to various parameters, such as the nature of the polymeric material (natural or synthetic), the structure (symmetric, asymmetric, porous, dense, etc.), the configuration (flat or cylindrical), or the physico-chemical properties of the final membrane (hydrophobic or hydrophilic). In the membrane separation processes, there are three basic forms of mass transport:

1. The "passive transport," in which the membrane acts as a physical barrier through which components are transported under the driving force of a gradient in their electrochemical potential.
2. The "facilitated" transport, the driving force for the transport of the various components is again the gradient in their electrochemical potential across the membrane. Here transport of a component is enhanced by the presence of a carrier.
3. The "active" transport, in which various components may be transported against the gradient of their electrochemical potential. This transport is possible when energy is provided to the system, for example by a chemical reaction within the membrane phase.

The transport of mass in a membrane is a nonequilibrium process and is conventionally described by the phenomenological equations called Fick's law, which relates the fluxes of matter to the corresponding driving forces. These driving forces are hydrostatic pressure, concentration, and electrical potential differences. For a given driving force, the flux through a unit membrane area is always inversely proportional to the thickness of the selective barrier [6]. Polymeric membranes share several similarities with cellular membranes: separation, compartmentalization, and selective transport of molecules. This is why even in the medical field a wide variety of polymeric membranes are highly important. Today, the membranes are used for blood oxygenation and purification in replacement or substitution of pulmonary and renal function and in hybrid artificial organs (bioartificial liver, bioartificial pancreas) for the therapeutic treatment of patients who are waiting for organ transplantation or the regeneration of partially damaged organs [3]. In bioartificial organs, membranes act as immunoselective barriers that serve to prevent the contact between the cells and the immunocompetent species present in the blood of the patient and at the same time permit the transport of nutrients and metabolites to and from the cells [7,8]. In such devices, the membrane also functions as a means for the oxygenation of the cells and as support for the anchorage-dependent cells like hepatocytes [9—11] mimicking in vitro the in vivo conditions and the cellular environment. For this reason, the characteristics of membranes like selective permeability, hydrophilicity/hydrophobicity, porosity, and elasticity can play a decisive role in the cell−membrane interaction and the induction of cell growth. Membranes can function also as scaffolds as 3D templates for initial cell attachment and subsequent tissue formation. Therefore, the choice of the membrane for a given application depends on its permeability characteristics, as well as its chemical and physical properties related to the separation process [12,13]. It has been proved, in fact, that cell morphology changes depending on the morphological, physico-chemical, and mechanical properties of the adhesion surface [4,14]. It is of primary importance that the membranes group composed by polymers the PEEK-WC-PU membranes, for example—combine the advantageous properties of both polymers (biocompatibility, mechanical strength, elasticity) with those of membranes like permeability, selectivity, and geometry. This membrane was able to support the long-term maintenance and differentiation of human hepatocytes in a biohybrid system [15].

5.2 Membranes for tissue engineering

Membranes can act as an instructive ECM for cells, especially for stem cells or progenitor cells, where differentiation is desired for their therapeutic potential and usefulness in toxicological testing. Like ECM, membrane exhibits from microscale to nanoscale of chemistry and topography and is able to provide physical, chemical, and mechanical signals to the cells, which are important for guiding their differentiation. It has been demonstrated that chitosan membrane, with its intrinsic properties such as its chemical similarity to the glycosaminoglycans of the liver ECM and the presence of protonated amino groups at physiological pH, allowed the expansion and functional differentiation of embryonic liver cells in hepatocytes (Figure 5.1) [16].

A big challenge is the use of membrane systems in the creation of systems in which neuronal cells are organized in a controlled manner. The cellular organization must include the control of the cell−cell interactions and of the surrounding environment, since the functionality and repair of neuronal cells depend on their intrinsic genetic program and the extracellular environment. Currently, tissue engineering of the nervous system involves (1) the functional replacement of one or more missing neuroactive components; (2) the rescue and/or regeneration of degenerated nervous tissue; (3) and the construction of fully functional neural circuits in vitro. Once the nervous system is compromised, recovery is difficult and malfunctions in other parts of the body may occur due to lack of cell division of mature neurons.

To enhance the axonal regeneration prospects and the functional recovery, the research has focused on designing "nerve guidance channels" or "nerve conduits." In particular, it has been demonstrated that hippocampal cells are susceptible to the characteristics of the membrane surface on which they are grown [1−17],

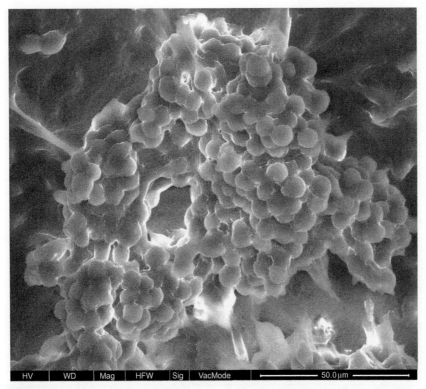

FIGURE 5.1

SEM's image of liver progenitor cells cultured on chitosan membranes.

while the micro-geometry of the surface and the molecular weight cutoff greatly influence the axonal regeneration. Also, the hollow fiber membranes have been used for biomedical applications and engineering, since they allow an efficient exchange of nutrients and metabolites to maintain the viability and cellular functions in vitro. In the neuronal regeneration context, the surface micro-geometry, the chemical nature of polymer and the membranes' molecular weight cutoff, the electrical properties, and the rate of degradation of materials [18] all appear to have a marked influence on the tissue response. It has been shown that neuronal outgrowth improves on polyacrylonitrile hollow fiber membranes with hydraulic permeance of 0.215 L/h $\times$ m^2, which enhances mass transfer of nutrients and metabolites to the cells and the removal of catabolites (Figure 5.2) [19]. Furthermore, hippocampal neurons express a different morphology in response to different surface properties of polymeric membranes. Both mechanical and chemical regulation are important for the growth and differentiation processes of neuronal cells in biohybrid systems: these substrate characteristics induce specific cellular

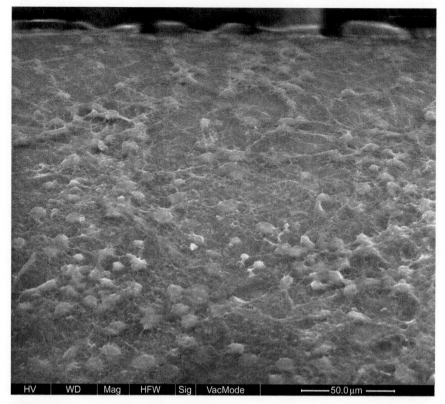

| HV | WD | Mag | HFW | Sig | VacMode | ├──────50.0 µm──────┤ |

FIGURE 5.2

Hippocampal neuronal cells on polyacrilonitrile HF membranes after 16 days of culture.

responses, allowing neurons to assume a definite orientation in space with the creation of a network of synaptic connections in an in vitro system. Neural cells cultured on smooth fluorocarbon (FC) and polyethersulfone (PES) membranes showed an important increase of neuritis outgrowth and spreading, whereas the use of membranes of modified polyetheretherketone (PEEK-WC), with greater roughness, induces cellular aggregation and processes extending into the pores of the membrane [1].

Bone tissue engineering typically involves the use of porous, bioresorbable scaffolds to serve as temporary, three-dimensional scaffolds to guide cell attachment, differentiation, proliferation, and subsequent tissue regeneration. Recent research strongly suggests that the choice of scaffold material and its internal porous architecture significantly affect regenerated tissue type, structure, and function [20]. The effects of mean pore size have been extensively studied [21,22], and Chang et al. showed that the direction of bone in growth was along the long axis of the porous channels [22]. In addition to possessing the appropriate material composition and internal pore architecture for regenerating a specific target tissue, scaffolds must have mechanical properties appropriate to support the newly formed tissue [23]. Conventional single-component polymer materials cannot satisfy these requirements. In fact, although various polymeric materials are available and have been investigated for tissue engineering, no single biodegradable polymer can meet all the requirements for bone engineering applications. Therefore, the design and preparation of multicomponent polymer systems represent a viable strategy in order to develop innovative multifunctional biomaterials. In particular, polymer−ceramic composites have been developed to combine the intrinsic properties of each component and to optimize the physicochemical and biological properties that the hard tissues need [24].

Traditional biodegradable polyesters used in the biomedical field, such as PLA, PGA, or PCL, are easily reabsorbed, show ductile properties, and can be processed to different devices. Bioceramics such as hydroxyapatite (HA) and tricalcium phosphate (TCP) have been shown to induce a good response from bony cells and have often been combined with biodegradable polymers to produce bone substitutes because of their structural similarity to the mineral phase of bone and their osteoconductive and bone binding properties. Indeed, hydroxyapatite $(Ca_5(PO4)_3(OH))_n$ is the main mineral component of bone, and because of its high bioactivity and biocompatibility, it is commonly used as a filler in polymer-based bone substitutes [25]. This approach also offsets the problems of brittleness and the difficulty of shaping hard ceramic materials to fit bone defects [26]. In fact, they are currently in use in orthopedics, despite their fragility, scarce remodeling, low flexibility, and moldability. A combination of both material types thus reduces their drawbacks while benefiting from their respective advantages [26]. When bioactive ceramics are hybridized with biodegradable polymers, the composite systems possess a level of flexibility, appropriate mechanical properties, and improved biological activity and osteoconductivity [27−29].

5.3 Hollow fiber polymeric membranes for cell and tissue perfusion

An efficient exchange of nutrients and metabolites to maintain the viability and cellular functions in vitro is really a big challenge for each in vitro regenerative process. Hence, the greatest innovation in the engineering and biomedical field is represented by the hollow fiber membranes. Their particular three-dimensional and porous structure allows the passage of molecular species that are smaller in size than their pores exclusively, so enable two solutions: one containing the components that can permeate the walls of the membrane, called the "permeate" and one constituted by larger molecules that are not able to pass through defined but instead "retentate." The membranes are then classified according to the molecular cut, or molecular weight cutoff, that is defined as the molecular weight of the species that is 90 percent retained by the membrane. Membranes with molecular weight cutoff in the range of 50,000 to 100,000 Da, in fact, have been used as immunoselective barriers to prevent the passage of immunocompetent species that are present in the blood of patients undergoing transplantation. However, the best application in recent years has been their use for the design of bioreactors, for the following reasons:

1. They offer the cells a three-dimensional structure on which to build a fabric.
2. They compartmentalize the cells in a well-controlled microenvironment.
3. They minimize the surface/volume ratio—that is, equal volume has a larger surface area available for cell adhesion and the exchange of metabolites, compared with a flat membrane.
4. They allow culture at high densities.

In particular, membranes in tubular configurations act as guidance channels to protect the lumen of regenerating axons from the external environment. Surprisingly, among the variety of tubular structures used for guidance channels, semipermeable hollow fiber membranes appear to be considered a favorable guide for the regeneration of neuronal tissue [30,31].

The permeability properties of the membranes may affect the axonal outgrowth by governing the mass transfer of nutrients, metabolites, and waste products between the shell of the membranes where the cells are cultured and the channel lumen environment where the medium is present.

The HF membranes could be more advantageous with respect to the flat membranes because they offer a wide adhesion area for neurons in a small volume, as well as a compartmentalization of cells in a physical space separated by the extracellular environment and a selective transport of molecules between the medium and cellular compartment through the porous structure of the membranes.

5.4 Membrane bioreactors for tissue constructs

The in vitro cultivation of three-dimensional constructs that support an efficient nutrition of the cells, combined with the application of a direct mechanical stimulation, affects cellular activity, differentiation, and specific functionality. Today, a wide variety of bioreactors have been developed for the engineering of tissues such as retinas, skin, muscles, ligaments, tendons, bones, cartilage, and liver. Ideally, a bioreactor should make it possible to control considerable environmental factors (pH, O_2, temperature, nutrient transfer, waste removal, etc.) at well-defined levels and allow operations like sampling and feeding, while avoiding the contamination that might normally occur in a traditional culture. Oxygenation levels in particular have proved to be critical for the matrix components' production in the cultivation of cartilage cells, despite the controversy about the benefits of a high or low oxygen concentration. It's now known that the mechanical forces improve or accelerate tissue regeneration and cell growth. The fluid dynamic stress, or shear stress, induced by the fluid flow through the surface of the construct and the open pore space is believed to be the most important mechanical stimulus for the activation of mechanic transductional signals [32].

Several membrane bioreactors have been developed on the basis of cell type. Beginning in 1999, when Lamers showed that cytokines can be produced by T lymphocytes using two parallel cellulose acetate (CA) hollow fiber membrane bioreactors, progress in the development of that kind of device has increased rapidly [33]. Jasmund et al. proposed a modified QUADROX oxygenator for the oxygenation of blood and the removal of the carbon dioxide [34]. They proposed fibers of polyethylene (PE) suitable for the exchange of heat and O_2. Thereafter, a complete purification of blood samples was obtained by modifying a hollow fiber membrane of CA with copolymers of MPC (2-metacryloilossiethylfosforilcolina): PMB 30 and the PMA 30 (MPC-co-n-butyl methacrylate, and MPC-co-methacrylic acid) [35].

Mimicking the in vivo organization of hepatocytes in plates separated by sinusoids, an oxygen-permeable membrane bioreactor was developed on the basis of the oxygen liver cell requirements [36]. Oxygen is one of the most important nutrients for cell viability, and its transport from atmosphere to the surface of cells is often limited because of its poor solubility in the culture medium. Oxygen is generally consumed as glucose at approximately the same rate but the oxygen solubility is lower (0.2 mM at atmosphere pressure) with respect to the glucose (20 mM). Due to the high O_2 dependence of primary hepatocytes in long-term cultures, oxygenating supports offer beneficial culture conditions, similar to physiologic ones (Figure 5.3). The bioreactor provides a culture chamber for cells delimited downstairs and upstairs by thick flat-sheet fluorocarbon membranes that are permeable to O_2, CO_2, and H_2O vapors. In this way, the membranes simulate the sinusoid functions, ensuring an optimal exchange of O_2 and CO_2 at the cellular level. Cells receive oxygen directly from the bottom membrane sheet, where they are seeded, and indirectly from the top membrane sheet in contact with the medium overlying the cells. Hepatocytes grew directly on the border between the gas and the liquid phases, where the culture medium did not act as a diffusion barrier, simulating the in vivo liver sinusoidal organization. The connection to a perfusion system, including a peristaltic pump, allowed the feeding of nutrients and metabolites and the continuous mixing of the medium inside the bioreactor. Human hepatocytes reconstituted many of the features of the in vivo liver, exhibiting a polyhedral shape and a three-dimensional structure with distinct pericellular zones and the formation of aggregates in some areas of the membrane.

The most commonly used membrane bioreactors in tissue engineering take advantage of tubular configuration, using material hollow fiber (HF) membranes as scaffolding. The tubular configuration of materials meets the main requirements for cell culture, such as the wide area for cell adhesion and the free permeation of oxygen, nutrients, and catabolites to and from cells, with the advantage

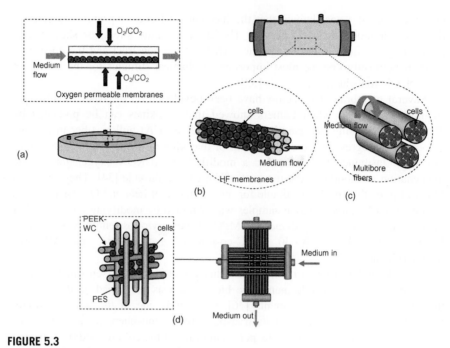

FIGURE 5.3

Membrane bioreactors by using membranes with different functions and configurations. (a) Oxygen permeable bioreactor. (b) Multibore fiber bioreactor. (c) Parallel fiber bioreactor. (d) Crossed HF membrane.

of compartmentalizing and protecting cells from shear stress and from the patient's plasma immunoreactive components. HF configuration may further orchestrate the initiating tissue formation by guiding the microarchitecture of the developing tissue by providing histo-appropriate mechanical forces and mechanical strength. Critical issues in the HF membrane bioreactors are the bioreactor configuration, the fluid dynamics, and the membrane properties that affect cell adhesion and mass transport. Besides the removal of catabolites (e.g., lactate, ammonium, carbon dioxide), the accumulation of toxins and toxic metabolites must be avoided in order to maintain viable and functional cells inside the bioreactor environment. Furthermore, the delivery of plasma proteins and large MW proteins (e.g., clotting factors) and drug metabolites must be ensured in both clinical and in vitro devices.

Mass transfer across the membrane occurs by diffusion and/or convention in response to existing transmembrane concentration or pressure gradients. Both mechanisms of transport should be taken into account in the design of HF membrane bioreactors [12]. Mass transfer to and from cells for nutrient supply and waste elimination, respectively, is a critical issue in any bioreactor design, especially when cells are cultured in three-dimensional multicellular aggregates, where mass-transfer limitation of oxygen and metabolism may occur in the core of aggregates [37]. Mixing HF bioreactors with different fibers or coaxial HF membrane reactors, which some of the authors have done, confirmed the importance of enhancing mass transfer [38]. Many kinds of HF membrane bioreactors with different configurations were described in the literature in which cells were compartmentalized in the lumen or in the extracapillary space of the HF membranes.

A completely new bioreactor in tubular configuration was developed by using multibore fiber membranes of modified polyethersulfone, in which each multibore fiber contains seven compartments represented by seven capillaries with diameters of 960 μm and an inner layer that is a very thin selective surface [39]. The capillaries are surrounded by a porous, foamy support structure that gives good stability and mechanical resistance to the fiber, minimizing the risk of membrane rupture. At the same time, the porous, foamy support structure ensures a high hydraulic permeance $(1.067 \, \text{L} \cdot \text{m}^{-2} \cdot \text{h} \cdot \text{mbar})$ and a permeability that is 1000 times higher than the filtration surface and higher than those of other membranes with similar pore size. The main advantage of using the multibore fibers is the presence of more cell compartments for each single fiber that communicate with one another, providing the ability to compartmentalize different types of cells in the same fiber and to favor cell−cell communication through paracrine signals. Furthermore, the mechanical resistance for a single multibore fiber is improved with respect to a single capillary.

Important shortcomings in the use of HF membrane bioreactors related to the limitation to the mass transport of nutrients and metabolites were handled by arranging fibers in cell plates with sinusoidal structures located on both sides. A particularly interesting approach was undertaken by using a multicompartment bioreactor with four different capillary membranes, each one serving a specific function: plasma inflow, plasma outflow, decentralized oxygen supply and carbon dioxide removal, and sinusoidal endothelial co-culture [40]. This complex geometry was proven to ensure adequate perfusion of cells with medium/plasma, oxygen, and removal of catabolites and liver-specific products. In a similar way—and with an approach that reduces the complexity of the bioreactor analysis, thus obtaining a satisfactory control of the operational parameters, including fluid dynamics optimization and the system performance—De Bartolo et al. developed a crossed hollow-fiber membrane bioreactor by using only two different kinds of membranes [41]: PEEK-WC and PES HF membranes with different molecular weight cutoffs, physico-chemicals, and permeability properties. These two types of hollow-fiber membranes were cross-assembled in alternating and perpendicular positions at 250 μm distance from each other to establish three separate compartments: two intraluminal compartments within the PEEK-WC and PES fibers and an extraluminal compartment formed by the fiber network where cells were seeded. On the basis of their characteristics, HF membranes performed different functions: PEEK-WC HF membranes provided cells with an oxygenated medium containing nutrients and metabolites, whereas PES HF membranes removed catabolites and cell products from cell compartments, mimicking the in vivo arterious and venous blood vessels and therefore the high in vivo hepatic perfusion. The combination of these two fiber sets resulted in a high mass exchange between the medium and cell compartments through the cross-flow of culture medium. Furthermore, the geometry of crossed HF membranes, with a constant distance of 250 μm between each fiber, allows achieving a homogeneous and small size of cell aggregates, which facilitate mass transfer and hence the perfusion of cells in the core of the small clusters within the network of the fibers [37]. As a result of the adequate perfusion conditions, human hepatocytes maintained their metabolic functions for all investigated period.

5.5 Future perspectives

The potentiality of applying polymeric membrane systems for biofabrication of tissues and organs is growing. Interesting results have been obtained by using membrane systems for the expansion and differentiation of stem cells, perfusion of engineered constructs, and the in vitro realization of liver and neural tissues. The use of these systems can be exciting in helping to find nature's substitutes and to solve the pathogenesis of important human diseases or to select an optimal pharmaceutical treatment. The ultimate goal will be the development of tissue and organ homologues from stem cells and biomaterials that mimic extracellular matrix in vivo, providing not only the geometry and physico-chemical properties to maximize the migration, adhesion, and proliferation of cells but also the capability to sequester and release biological signals essential for cell molecules.

References

[1] De Bartolo L, Rende M, Morelli S, Giusi G, Salerno S, Piscioneri A, et al. J Memb Sci 2008;325:139—49.

[2] Kaihara S, Vacanti JP. Tissue engineering: toward new solutions for transplantation and reconstructive surgery. Arch Surg 1999;134(11):1184—8.

[3] Langer R, Vacanti JP. Tissue engineering. Science 1993;260:920—6.

[4] De Bartolo L, Morelli S, Giorno L, Campana C, Rende M, et al. J Memb Sci 2006;278:133—43.

[5] Chapekar MS. Tissue engineering: challenges and opportunities. J Biomed Mater Res 2000;53(6):617—20.

[6] Porter MC. Handbook of industrial membrane technology. Noyes Publication; 1990.

[7] Drioli E, De Bartolo L. Membrane bioreactor for cell tissue and organoids. Artif Organs 2006;30:793—802.

[8] Morelli S, Salerno S, Piscioneri A, Campana C, Drioli E, et al. Membrane bioreactors for regenerative medicine: an example of the bioartificial liver. Asia-Pacific J Chem Engg 2010;5(1):146—59.

[9] De Bartolo L, Jarosch-Von Schweder G, Haverich A, Bader A. A novel full-scale flat membrane bioreactor utilizing porcine hepatocytes: cell viability and tissue specific functions. Biotech Progress 2000;16:102−8.

[10] Bader A, De Bartolo L, Haverich A. High level benzodiazepine and ammonia clearance by flat membrane bioreactors with porcine liver cells. J. Biotechnology 2000;81:95−105.

[11] De Bartolo L, Morelli S, Rende M, Gordano A, Drioli E. New modified polyether-etherketone membrane for liver cell culture in biohybrid systems: adhesion and specific functions of isolated hepatocytes. Biomaterials 2004;25:3621−9.

[12] Curcio E, De Bartolo L, Barbieri G, Rende M, Giorno L, et al. Diffusive and convective transport through hollow fiber membranes for liver cell culture. J Biotechnol 2005;117:309−21.

[13] Unger RE, Huang Q, Peters K, Protzer D, Paul D, et al. Growth of human cells on polyethersulfone (PES) hollow fiber membranes. Biomaterials 2005;26:1877−84.

[14] De Bartolo L, Morelli S, Bader A, Drioli E. Evaluation of cell behaviour related to physico-chemical properties of polymeric membranes to be used in bioartificial organs. Biomaterials 2002;23(12):2485−97.

[15] De Bartolo L, Morelli S, Gallo MC, Campana C, Statti G, et al. Effect of isoliquiritigenin on viability and differentiated functions of human hepatocytes on PEEK-WC-polyurethane membranes. Biomaterials 2005;26:6625−34.

[16] Piscioneri A, Campana C, Salerno S, Morelli S, Bader A, et al. Biodegradable and synthetic membranes for the expansion and functional differentiation of rat embryonic liver cells. Acta Biomater 2011;7:171−9.

[17] Morelli S, Salerno S, Piscioneri A, Papenburg BJ, Di Vito A, et al. Influence of micropatterned PLLA membranes on the outgrowth and orientation of hippocampal neuritis. Biomaterials 2010;31:7000−11.

[18] Maquet V, Martin D, Scholtes S, Franzen R, Schoenen JJ, et al. Poly(D,L-lactide) foams modified by poly(ethylene oxide)-block-poly(D,L-lactide) copolymers and a-FGF: in vitro and in vivo evaluation for spinal cord regeneration. Biomaterials 2001;22(10):1137−46.

[19] Morelli S, Piscioneri A, Salerno S, Rende M, Campana C, et al. Flat and tubular membrane systems for the reconstruction of hippocampal neuronal network. J Tissue Eng Regen Med 2012;6:299−313.

[20] Karageorgiou V, Kaplan D. Porosity of 3D biomaterial scaffolds and osteogenesis. Biomaterials 2005;26(27):5474−91.

[21] Zeltinger J, Sherwood JK, Graham DA, Mueller R, Griffith LG. Effect of pore size and void fraction on cellular adhesion, proliferation, and matrix deposition. Tissue Eng 2001;7:557−72.

[22] Chang BS, Lee CK, Hong KS, Youn HJ, Ryu HS, et al. Osteoconduction at porous hydroxyapatite with various pore configurations. Biomaterials 2000;21(12):1291−8.

[23] Hutmacher DW. Scaffold design and fabrication technologies for engineering tissues—state of the art and future perspectives. J Biomater Sci Polym Ed 2001;12:107−24.

[24] Wang M. Developing bioactive composite materials for tissue replacement. Biomaterials 2003;24:2133−51.

[25] Azevedo MC, Reis RL, Claase MB, Grijpma DW, Feijen J. Development and properties of polycaprolactone/hydroxyapatite composite biomaterials. J. Mater Science: Mater Med 2003;14:103−7.

[26] Zhao J, Guo LY, Yang XB, Weng J. Preparation of bioactive porous HA/PCL composite scaffolds. Appl Surf Sci 2008;255:2942—6.

[27] Shor L, Güçeri S, Wen X, Gandhi M, Sun W. Fabrication of three-dimensional polycaprolactone/hydroxyapatite tissue scaffolds and osteoblast-scaffold interactions in vitro. Biomaterials 2007;28:5291—7.

[28] Lebourg M, Suay Antón J, Gomez Ribelles JL. Characterization of calcium phosphate layers grown on polycaprolactone for tissue engineering purposes. Comps Sci Tech 2010;70:1796—804.

[29] Kim HW, Knowles JC, Kim HE. Porous scaffolds of gelatin—hydroxyapatite nanocomposites obtained by biomimetic approach: characterization and antibiotic drug release. J Biomed Mater Res B 2004;70B:240—9.

[30] He Q, Zhang T, Yang Y, et al. In vitro biocompatibility of chitosan-based materials to primary culture of hippocampal neurons. J Mater Sci: Mater Med 2009;20:1457—66.

[31] Zhang N, Zhang C, Wen XJ. Fabrication of semipermeable hollow fiber membranes with highly aligned texture for nerve guidance. Biomed Mater Res 2005;75A:941—9.

[32] Vance J, Galley S, Liu DF, Donahue SW, Vance J, et al. Mechanical stimulation of MC3T3 osteoblastic cells in a bone tissue-engineering bioreactor enhances prostaglandin E2 release. Tissue Eng 2005;11:1832—9.

[33] Lamers CHJ, Gartama JW, Luider-Vrieling B, Bolhuis RLH, Bast EJEG. Large-scale production of natural cytokines during activation and expansion of human T lymphocytes in HF bioreactor cultures. J. Immunother 1999;22(4):299—307.

[34] Jasmund L, Langsch A, Simmoteit R, Bader A. Cultivation of primary porcine hepatocytes in an OXY-HFB for use as a bioartificial liver device. Biotechnol Prog 2002;18:839—46.

[35] Ye SH, Watanabe J, Takai M, Iwasaki Y, Ishihara K. High functional hollow fiber membrane modified with phospholipid polymers for a liver assist bioreactor. Biomaterials 2006;27:1955—62.

[36] De Bartolo L, Salerno S, Morelli S, Giorno L, Rende M, et al. Long-term maintenance of human hepatocyte in oxygen-permeable membrane bioreactor. Biomaterials 2006;27(27):4794—803.

[37] Curcio E, Salerno S, Barbieri G, De Bartolo L, Drioli E, et al. Mass transfer and metabolic reactions in hepatocyte spheroids cultured in rotating wall gas-permeable membrane system. Biomaterials 2007;28:5487—97.

[38] Gerlach JC, Encke J, Hole O, Muller C, Ryan CJ, et al. Bioreactor for a larger scale hepatocyte in vitro perfusion. Transplantation 1994;58:984—8.

[39] De Bartolo L, Morelli S, Rende M, Campana C, Salerno S, et al. Human hepatocyte morphology and functions in a multibore fiber bioreactor. Macromol Biosci 2007;7:671—80.

[40] Gerlach J, Trost T, Ryan CJ, Meissler M, Hole O, et al. Hybrid liver support system in a short term application on hepatectomized pigs. Int J Artif Organs 1994;17:549—53.

[41] De Bartolo L, Salerno S, Curcio E, Piscioneri A, Rende M, et al. Human hepatocyte functions in a crossed hollow fiber membrane bioreactor. Biomaterials 2009;30:2531—43.

Laser-Assisted Bioprinting for Tissue Engineering

Bertrand Guillotin, Muhammad Ali, Alexandre Ducom, Sylvain Catros, Virginie Keriquel, Agnès Souquet, Murielle Remy, Jean-Christophe Fricain, Fabien Guillemot

INSERM U1026, Bordeaux, F-33076 France; Univ. Bordeaux Segalen, Bordeaux, F-33076 France

CONTENTS

INTRODUCTION

The loss or failure of an organ or tissue is one of the most frequent, devastating, and costly problems in health care [1]. Current treatment modalities include transplantation of organs, surgical reconstruction, use of mechanical devices, and supplementation of metabolic products. Epidemiological studies highlight tissue/organ shortages, which justify original approaches to fulfill clinical needs [2]. The architecture of a given tissue is characterized by its constituting cell types, the biochemical and mechanical properties of the extracellular matrix, and the topology of these components. Accordingly, tissue engineering (TE) aims at providing regenerative medicine with original products. In addition, TE may provide original models (2D/3D) for fundamental research in biology and pharmacokinetics studies [3−5]. Since the late 1980s and the creation of the first workable definition of hybrid artificial organs [6], an increasing number of research groups throughout the world have developed TE-orientated approaches. As stated by Langer and Vacanti [7], these approaches apply the principles of engineering and life sciences to the development of biological substitutes that restore, maintain, or improve tissue or whole organ function. Generating biological tissues in vitro involves the use of engineering and material methods, the appropriate combination of cells, and the suitable biochemical and physicochemical factors to mimic both the microenvironment of cells and the microarchitecture of tissues in the body.

More precisely, live tissues and organs are composed of multiple cell types, which are assembled and interfaced at the micrometer scale. In the liver, for example, columns of hepatocytes are interfaced with biliary capillaries on the apical side and with sinusoidal blood vessels on the basal side to form lobules. Such high-density, compartmentalized, and integrated cellular organization has two major functional outcomes: homeostasis—in particular metabolic exchange—is optimized, and functional units are packed together to form organs with a physiological efficiency that is compatible with living tissues. Consequently, miniaturization of TE processes (i.e., microscale organization of cells) might be necessary to fabricate organotypic structures that compare favorably with the functionality of live organs.

Tissue engineering approaches can be divided into three strategies based on the level of spatial organization. First, macroscopic strategy can be likened to traditional TE in which cells are seeded onto a macroporous scaffold. Cells are expected to colonize the inner volume of the scaffold by cell mobility and proliferation and fluid flow. However, cell penetration and adhesion being challenged, several months may be required for the cells to adhere and proliferate through the scaffold. As a result, cellularized scaffolds do not present the ability to mimic the functional multicellular anisotropy and density of the host tissue. Second, mesostructures or modular blocks, also termed organoids [8,9,10], are based on the ability of the cells to self-assemble and their capacity to maintain viability and function when located within the diffusion limit of nutrient supply. The modular approach enables the production of 3D modules in a variety of shapes (e.g., cylinders), with a diameter between 40 and 1000 μm and cell densities of 10^5 to 10^8 cells/cm^2, and allows fabrication of multicellular constructs (e.g., bone-mimicking constructs including osteoblasts, osteoclasts, and endothelial cells). However, both macroscopic and mesoscopic approaches have not yet demonstrated the ability to mimic the functional multicellular anisotropy and density of the host tissue, in which cell-to-cell communication is critical for tissue morphogenesis, homeostasis, and healing. Consequently, it is also challenging to design an efficient perfusion system that is physiologically interfaced with the engineered tissue and that will branch to the host's vasculature at implantation. Reproducing the local cell microenvironment can be thought of as the ultimate target for TE and cell patterning. Conceptually, it could be addressed thanks to the capacity of positioning a single cell into its most suitable environment. Such a cell niche manufacturing approach is unique in its purpose of dealing with tissue complexity and engineering a desired tissue from the bottom up.

In parallel to scaffold-based approaches involving cell seeding of porous structures [11], some authors have suggested three-dimensional biological structures can be built from the bottom up by the technology of bioprinting: the automated, computer-aided deposition of cells, cell aggregates, and biomaterials [10,12–14]. Since the pioneering work performed by Klebe et al. [15,16], technological advances in the field of automation, miniaturization, and computer-assisted design/computer-assisted manufacturing (CAD/CAM) have accompanied the development of several bioprinting technologies. Inkjet printers have been successfully used to pattern biological assemblies according to a computer-aided design template [17–19]. Pressure-operated mechanical extruders have also been developed to handle living cells and cell aggregates [12]. Parallel to these methods, laser-assisted bioprinting technologies have emerged as alternative methods for the assembly and micropatterning of biomaterials and cells. Laser-assisted bioprinting of biological material in general, and living cells in particular, is based on the laser-induced forward-transfer (LIFT; Figures 6.1 and 6.2) technique in which a pulsed laser is used to induce the transfer of material from a source film [20–22]. Several declinations of LIFT have been implemented to print living cells [23,24]. For convenience, we will use the general term *laser-assisted bioprinting (LAB)*, although we will mainly discuss results obtained with setups that require an intermediate light-absorbing layer of metal. CAD/CAM workstations of LAB (Figure 6.3) have been designed and used to print viable cells and to organize them with cell-level resolution, with high throughput, in two-dimensional and three-dimensional tissue constructs, with the goal of mimicking the functional histology of live tissues. Thus, under suitable laser pulse conditions, and for liquids presenting a wide range of rheologies, the materials can be deposited in the form of well-defined circular droplets with a high degree of spatial resolution [25–27].

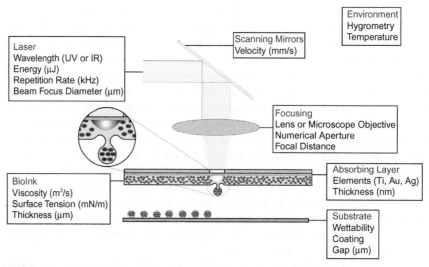

FIGURE 6.1

A typical LIFT experimental setup is generally composed of three elements: a pulsed laser source, a target or ribbon coated with the material to be transferred, and a receiving substrate. The ribbon is a three-layer component: a support that is transparent to the laser radiation wavelength, coated with a thin absorbing layer (50 nm); coated itself with a transfer layer (50 μm), named bioink, that contains the elements to be printed like biomaterials, cells, or biomolecules [32].

In the following sections, we first present LAB from a technical point of view, and then the physics behind the technology are detailed. Accordingly, we then focus on the physical parameters that must be tuned to print viable cell patterns with respect to cell-level spatial resolution in high-throughput conditions. Finally, we report on TE applications of LAB for basic research in biology as well as for regenerative medicine.

6.1 LAB's terms of reference
6.1.1 Workstation requirements

To design a LAB workstation, various pulsed lasers were evaluated for their suitability with living cells and biomaterials, as well as for rapid prototyping applications, all leading to comparable conclusions in terms of cell viability and printing resolution. The following major requirements were considered:

1. The wavelength λ should not induce alteration of biological materials. Due to the potential denaturation of DNA by UV lighting, and despite short pulse

(a)

(b)

(c)

FIGURE 6.2

(a) View of the high-throughput LAB. (b) Optomechanical setup. (c) High-resolution positioning system placed below the carousel holder with a loading capacity of five different ribbons.

(a)

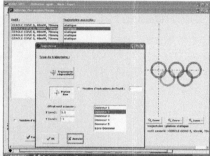

(b)

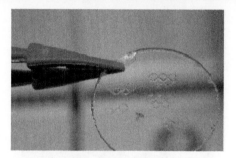

FIGURE 6.3

(a) Example of CAD/CAM software to help design the pattern of manufacturing as well as to pilot the laser beam accurately. (b) Printed pattern according to the CAD/CAM design.

duration and the presence of the light-absorbing metal layer, near infrared lasers might be preferable to UV lasers.

2. The pulse duration τ and the repetition rate f must be considered with the purpose of high-throughput processes.

3. The beam quality, including divergence (q), spatial mode, and pulse-to-pulse stability (ptp) has to be taken into account to ensure the reproducibility, stability, and high resolution of the system.

4. A laser-based workstation dedicated to tissue engineering applications should be designed with the purpose of executing various tasks rather than solely bioprinting. Consequently, the mean laser power should be high enough to perform additional processes like photolithography, photo polymerization, machining, sintering, and foaming [28–30].

5. The laser repetition rate (in combination with scanning velocity; see following) has to be taken into account to avoid coalescence of vapor bubbles within the thickness of the bioink and reciprocal perturbation of consecutive jets. For example, successive laser spots should not overlap to avoid an alteration of the energy conversed by the absorbing layer. Also, sufficient

distance should be considered to prevent one droplet to land on and displace the precedent droplet that has been printed onto the substrate. Hence, the spatio-temporal proximity between both laser pulses and ejected droplet has to be taken into account in order to obtain a high printing resolution and reproducible results.

6. Galvanometric mirrors are useful to drive the laser beam onto the ribbon. This approach offers a higher manufacturing speed (compared to a design where the laser beam is fixed and the sample is translated by a motorized stage [31]. However, a refractive lens is required to correct the incidence of the laser beam onto the ribbon [32].

7. By analogy with inkjet printer cartridges, several ribbons of different materials can be processed by LAB. We have designed a carousel wheel on which up to five elements can be loaded and presented under the laser beam automatically (see Figure 6.2c and [32]).

8. A dedicated CAD/CAM software interfaces with the workstation (see Figure 6.3).

6.1.2 Droplet ejection mechanism

A typical LAB experimental setup is generally composed of three elements: a pulsed laser source, a target coated with the material to be printed, and a receiving substrate (see Figure 6.1). As described in the literature [21,33−35], the generation of microdroplets by LIFT proceeds through six consecutive steps:

1. Laser energy is first deposited into the skin depth of the absorbing layer.
2. The absorbing layer is then heated in its skin depth.
3. This induces heating of a very thin film of the bioink near the absorbing layer.
4. Depending on the laser irradiation intensity, a vapor bubble is generated.
5. The vapor bubble expands.
6. This triggers the bioink−air interface deformation. Accordingly, three regimes are usually observed: subthreshold, jetting, and plume regimes (Figure 6.4).

The regimes are intimately related with the laser-induced bubble dynamics, which can be roughly approximated by the Rayleigh-Plesset equation [36,37]:

$$R\ddot{R} + \frac{3}{2}\dot{R}^2 = \frac{kP_l}{\rho_l}\left(\frac{R_0}{R}\right)^{3\gamma} - \frac{P_l - P_v}{\rho_l} - 4\upsilon\frac{\dot{R}}{R} - \frac{2\sigma_l}{\rho_l R} \qquad (6.1)$$

where R is the vapor bubble radius, ρ_l is the liquid density, P_l is the liquid pressure, υ is the kinematic viscosity, σ_l is the surface tension, γ is the ratio of specific heats, P_v is the saturated pressure in the bubble, and k is the ratio of gas pressure in the bubble, P_g, to the liquid pressure P_l. This equation is the expression of the vapor bubble radius evolution versus time. It mainly depends on the liquid kinematic viscosity and surface tension.

The Rayleigh-Plesset equation describes bubble dynamics in an infinite volume. Because the size of the vapor bubble is not negligible compared to the

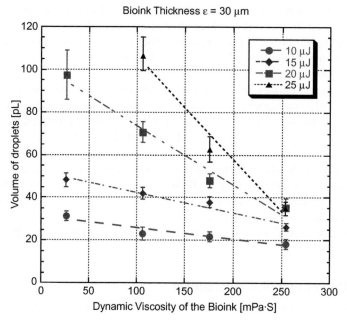

FIGURE 6.4

Evolution of droplet volume with dynamic viscosity for different laser pulse energy (10 to 25 μJ) using 30 μm bioink thick films, at a 500 μm printing distance (unpublished data). No droplet was collected onto substrates using 5 μJ laser pulse.

bioink thickness, the interactions of the bubble with the free surface have to be taken into account. Toward this end, Pearson et al. [38] and Robinson et al. [39] have demonstrated that (1) when the bubble reaches its maximum diameter, it begins to collapse under the external pressure strengths, and (2) a jet may be formed according to the dimensionless standoff distance

$$\Gamma = \frac{h}{R_{max}} \tag{6.2}$$

which is the ratio between the distance h (distance between the initial vapor bubble centroid and the free surface) and the maximum bubble radius, R_{max}. In Eq. (6.2), the maximum bubble radius, R_{max}, depends on laser energy, through the k ratio in the Rayleigh-Plesset equation, and viscosity of the bioink, while h is related to the initial thickness of the bioink film.

Recent results obtained by time-resolved imaging (Figure 6.5) have emphasized the importance of interactions between the bubble and the free surface into the jet formation by highlighting the existence of a critical vertex angle as a limiting boundary between subthreshold (Figure 6.5a) and jetting regimes (Figure 6.5b). Figure 6.6 shows the evolution of the vertex angle for subthreshold and jetting conditions as a function of time. In the subthreshold regime, the vertex angle is 105

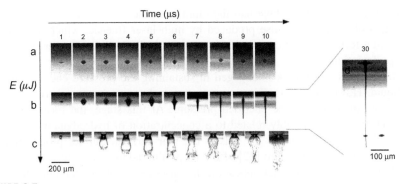

FIGURE 6.5

Time resolved imaging of laser-induced jet formation in LAB. (a) First row represents subthreshold regime, (b) second row denotes jetting regime, (c) third row shows plume regime, and (d) at the right, the process of depositing droplet in jetting regime is shown.

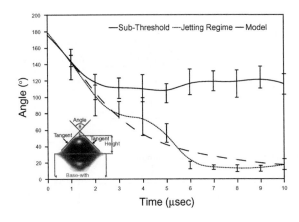

FIGURE 6.6

Vertex angle versus time. Solid curve denotes subthreshold regime, fine dot curve denotes jetting regime experimental data, and dashed curve denotes Dirichlet hyperboloid model [67].

degrees at 5 μsec, which increases subsequently as the protrusion recoils back. The initial rapid decrease in the vertex angle is attributed to creation of a high-pressure vapor bubble due to plasma formation from ablation of the gold layer [40,41]. Laser ablation produces a strong impulse, so pressure enclosed in the bubble surpasses the atmospheric pressure and surface energy. During the first 5 μsec, the expanding bubble deforms the bioink by stretching it in a forward direction. At 5 μsec, the influence of surface tension is apparent from the uniform round tip of protrusion. At 105 degrees in the subthreshold regime, the surface tension and viscoelastic properties of the ribbon have managed to counter the expansion of the bubble, and consequently the bubble retreats without producing a jet. In the case of jetting, the vertex angle drops with an abrupt change in behavior at 4 μsec. The initial expansion of the bubble and the rapid decrease in the vertex angle are the indicators of a strong impulse produced by high laser fluence, which accelerates bioink in a forward direction, helping it to cross the 105° limit to emerge as a jet. Abrupt deviation from the model at 4 μsec along with the appearance of a spike, which indicates that bioink is engaged in sustaining the pressure enclosed inside the liquid. However, its viscoelastic properties and external atmospheric pressure are unable to overcome the enclosed pressure. At this stage, the formation of a spike on the axis is possibly related to a concentration of hydrodynamic pressure [35] at the singular point on the axis at the rear surface. Further propagation is maintained by the momentum [42] of the jet pulling additional fluid from the surrounding film. The counterjet appears as a consequence of momentum conservation. Moreover, induction of fluid from the surrounding film in the direction of the flow increases jet inertia, which helps the protrusion to suppress the recoiling process due to surface tension [41].

Consequently, the three regimes generally observed in LIFT experiments (Figure 6.7) are not solely the result of laser irradiation conditions but also of

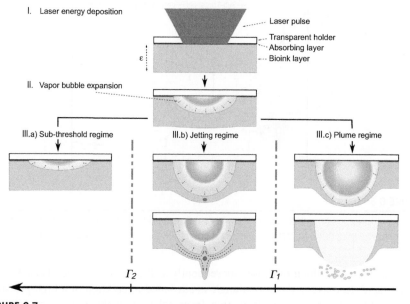

FIGURE 6.7

Mechanism for laser-induced droplet ejection. A vapor bubble is generated by vaporization of the absorbing layer and/or the first molecular layers of the liquid film. At given bioink viscosity and film thickness, jetting is observed for intermediary values of laser fluences: $\Gamma_1 < \Gamma < \Gamma_2$ (see IIIb). For a lower fluence $\Gamma < \Gamma_2$, the bubble collapses far from the free surface without generating a jet (see IIIa). For a higher fluence $\Gamma < \Gamma_1$, the bubble bursts to the surface, generating submicrometer droplets (see IIIc). Increasing film thickness or bioink viscosity leads to increased threshold Γ values. The three regimes are revealed by umbroscopy in time resolved imaging (unpublished data).

rheological properties and film thickness of the bioink. In other words, jetting is not simply occurring on the basis of an energy threshold mechanism [34] but rather on the basis of a complex Γ (E, ε, υ) threshold mechanism. Over a given laser energy for which a vapor bubble is formed at the absorbing layer—bioink interface, the three above-mentioned regimes can be distinguished (Figure 6.7):

- If Γ is higher than a threshold value Γ_1: the droplet ejection cannot occur, unless the substrate is close to the target. These are the conditions to observe the subthreshold regime (Figure 6.7a).
- When Γ is lower than a threshold value Γ_2: the bubble expands until it bursts, giving rise to the plume regime (Figure 6.7b).
- If Γ is ranged between Γ_1 and Γ_2: the bubble expands, then collapses, and finally a jet is formed (Figure 6.7c).

6.2 LAB parameters for cell printing
6.2.1 Bioink composition considerations

LAB requires cells to be suspended in a liquid bioink prior to being printed onto the substrate. Also, in order to print cell-laden 3D constructs, the bioink should gel after printing onto the substrate as fast as possible or be crosslinked by UV or chemicals, for example. The gelling and/or crosslinking process is necessary to stabilize the printed 2D pattern and to support the subsequent ink layer for 3D constructs using the layer-by-layer approach. This process should not be harmful to the cells. In addition to these properties that are required for jet-based technologies, the bioink should harbor properties reminiscent of physiological extracellular matrix (ECM), which is critical for cell homeostasis in vivo [43,44]. According to these terms of reference, the cells have been successfully printed by LAB using various solutions like culture medium alone [45] or in combination with sodium alginate [46] or thrombin [27]. Another concern is related to the evaporation of the bioink film that is typically served on the ribbon as a 50-μm thin layer. Accordingly, cell printing should be performed under a controlled atmosphere in terms of hygrometry (such conditions are also desired when biological material is processed). Othon et al. [47] have proposed supplementing the bioink with methylcellulose to help keep the bioink film from drying.

6.2.2 Viscosity of the bioink and laser energy influence cell viability

Printed cells may not retain viability for any conditions of LAB. It has been previously shown and numerically modeled that a minimum shock-absorbing mattress of hydrogel, like Matrigel^{TM}, was a required receiving substrate to absorb the mechanical shock of the printed cells [48,49]. Increasing the viscosity of the bioink, using sodium alginate, could improve cell viability where the mattress thickness is not sufficient to absorb the mechanical shock [50]. Although the viscosity of the bioink is critical, the laser energy deposit should also be reduced to print viable cells on a thin film of shock-absorbing substrate. No LAB-induced alteration of cell biology (in terms of phenotype and DNA knicks) has been detected so far [47,51,52,53], which validates de facto LAB for engineering cell-containing tissues for basic research. However, further studies should be implemented to rule out any genotoxicity of LAB, in cases of using LAB for cell-based clinical application.

6.3 High-resolution/high-throughput trade-offs

6.3.1 Tissue engineering needs versus LAB limits

For organ printing applications of LAB, cells should be printed according to a density and an organization that compares favorably to a parenchyma—that is, a tissue essentially made of cells, with virtually no ECM, high histological organization with rich blood perfusion, and multiple cell types. Assuming that LAB is able to print cells one by one and next to one another, with each cell having a diameter of 10 μm, it would require 28 hours to print a volume of 1 cm^3 with single-cell resolution at a printing frequency of 10 kHz. In other words, for a given volume to fabricate, given a constant printing speed, increasing the organization of the said volume would be more time consuming. Particular attention should then be brought to the feasibility to printing highly complex cell patterns at a high printing speed. We have previously demonstrated that high-throughput printing by LAB was achievable [14,27]. Concerning cell printing specifically, we may have reached a throughput threshold of around 5 kHz (Figure 6.8). Indeed, there are some spatiotemporal limits to jet formation and ejection due to fluid dynamics that limit resolution (Figure 6.9). The time of ejection of a jet is on the order of several tens of microseconds, which compares to the 5 kHz frequency that we classically use. The LIFT model strongly suggests that the jet produces some draining effect at the surface of the bioink film [35] that most likely accounts for the perturbation of the subsequent jet. Both pulse frequency and draining effect imply that two consecutive jets cannot be fired adjacent to each other above a certain firing frequency, which means that the laser beam should be driven fast enough (by galvanometric mirror) or the droplet receiving substrate holder should be moved fast enough between two consecutive pulses so that the subsequent jet occurs where the bioink film has not been disturbed by the precedent jet. According to these physical limits, LAB printing speed is theoretically merely limited by the highest laser pulse frequency available, according to the fastest galvanometric mirror available, providing the fact that intact bioink film is presented under the laser beam at the same time.

6.3.2 Printing resolution

Printing resolution can be defined as the number of printed droplets on a given length. This implies that the highest resolution is obtained by printing droplets as

FIGURE 6.8

Spatio-temporal limit of cell printing by LAB. Cells were printed onto glass according to five parallel lines of varying scanning speed (from top to bottom): 100, 200, 400, 800, and 1600 mm/s (laser pulse frequency of 5 kHz). The highest printing resolution of cells is achieved by using a scanning speed of 400 mm/s. With lower scanning speeds, the precision of droplet deposit is impaired, likely due to the overlap of successive jets.

small as possible and adjacent to one another. The high printing resolution of liquids achieved by LAB has been demonstrated (Figure 6.10 and [54,55,56]). The size of the droplet depends on the ink viscosity and the laser energy deposit. The laser energy deposit can be modulated by tuning the energy of the laser source and/or by cutting the laser beam with a diaphragm aperture stop. The higher the viscosity and/or the lesser the energy, the smaller the droplet diameter. It is possible to achieve similar resolution—that is, similar droplet size—with a 0.1% (w/v) alginate ink printed with a laser energy of 6 μJ (droplet size: (49 ± 3.5) μm, n = 15), and with a 1% alginate (w/v) ink printed with a pulse energy of 12 μJ (droplet size: $[51 \pm 4.2]$ μm, n = 15) [27]. By adapting the energy to a given viscosity, a wide range of ECM characterized by as many different viscosities can thus be printed at a similar resolution.

6.3.3 High cell density printing with cell-level resolution

To achieve microscale cell printing precision (see Figure 6.8), cells should be printed with a minimal volume of surrounding ECM (or bioink). However, because LAB is a LIFT-based and nozzle-free device, the number of cells in each ejected droplet is statistic [25]. The LAB nozzle-free setup precludes the cell printing process from clogging issues. Thus, bioink loaded with cell densities

(a)

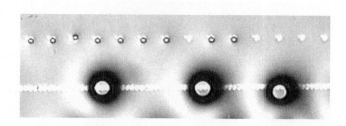

(b)

ribbon

air gap: 300 μm

substrate

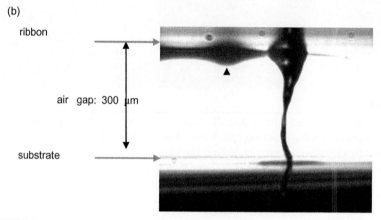

FIGURE 6.9

Spatio-temporal limit of jet formation (unpublished data). (a) The ribbon is observed after the bioprinting of a solution of sodium alginate 1% at a speed of 5 kHz. Upper line: One bubble per spot is visible when the laser is moved at 200 mm/s. The spots are 200 μm apart. Lower line: Bubbles coalesce to form larger bubbles when the laser is moved at 50 mm/s. The spots are contiguous, magnification 25×. (b) Two consecutive jets are observed by umbroscopy (unpublished data). The laser beam is moved from left to right. The first jet (pointed by the arrowhead) is retracting into the film of bioink, while the following jet is observed. Contact of the jet with the receiving substrate can be seen. Distance between the ribbon and the substrate is 300 μm. The two consecutive jets overlap partially due to draining/capillarity effect.

comparable to cell confluence observed in living tissue can be used. The presence of a single cell in each printed droplet is challenged by the use of a bioink with a low cell concentration—for example, 5×10^7 cells/mL. If cell density is too low on the ribbon, the ejected droplet of ink may not contain any cells [27]. To overcome this problem, at least two strategies can be proposed: First, increasing the laser energy deposit leads to the ejection of bigger droplets. As a result, cells are more likely to be dragged off by a draining/capillary effect. Second, cell density can be increased up to the point where the cells are touching one another at the

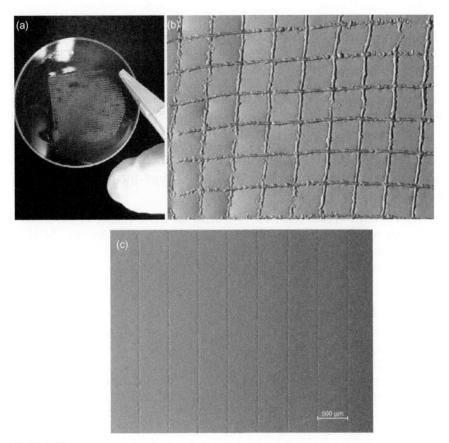

FIGURE 6.10

Rapid prototyping of a soft free-form hydrogel. The bioink composed of thrombin (250 UI) and CaCl₂ (40 mM) has been printed onto a layer of fibrinogen (90 mg/mL), according to a computer-designed matrix of orthogonal lines (2 cm length, 500 μm pitch) at a speed of 5 kHz and 200 mm/s. (a) Photograph of the fibrin pattern, scale 1:1. (b) Fibrin pattern observed with a phase contrast optical microscope, magnification 25×. (c) Parallel lines of printed Matrigel™ (unpublished data).

surface of the ribbon—that is, 1×10^8 cells/mL. If the droplet diameter is large enough, two contiguous printed droplets may coalesce, thus drawing a continuous line of cells or material (see Figure 6.10). Although increasing the size of the droplet leads to printing more cells at a time, the counterpart is a decrease in cell printing resolution. If single-cell printing resolution is desired—that is, one-by-one, next to one another—then the smallest droplet should be ejected using the lowest possible laser energy above the cell printing threshold and implying that (1) cell density is high enough so there is always one cell in the field of the laser beam, or (2) further development may focus on the implementation of a cell recognition scanning technology onto the ribbon prior to printing so the laser beam could exactly aim one single cell per pulse [57]. Accordingly, LAB can print cells virtually one by one from a high cell concentration bioink (1×10^8 cells/mL) to fabricate a tissue engineered product with comparable organization and cell density with living tissues in which multiple cell types are in physical contact with one another.

6.4 Three-dimensional printing: the layer-by-layer approach

The layer-by-layer approach aims at organizing the entire volume of the 3D construct with biological material like cells and proteins that form the ECM to achieve histological complexity and function (Figure 6.11 and [10,58]). In the following sections we present strategies to build up 3D constructs.

6.4.1 The bioink as a scaffold

As we saw in the last section, the printable bioink may be designed to provide not only biochemical elements but also some physical support for subsequent layers. For instance, a 3D structure was achieved with ADSC in a mixture of blood plasma and sodium alginate [59,60]. The final construct was then solidified by spraying the CaCl$_2$ crosslinker. The same team successfully stacked two arrays of spatially organized ECFC and ADSC using a bioink composed of fibrinogen and hyaluronic acid [60]. Each layer of array was wetted in or sprayed with thrombin solution, resulting in

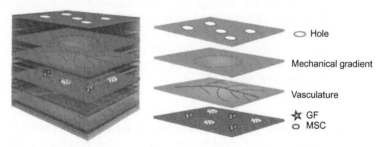

FIGURE 6.11

Schematic principle of the layer-by-layer assembly of complex tissue constructs. These complex tissues feature micropatterns of cells like mesenchymal stem cells (MSC), biochemical cues like growth factors (GF), physical cues like stiffness gradients, and defined shapes like holes. Holes can be processed by laser micro-machining to favor fluid diffusion through the structure and could also be endothelialized.

a stable construct made of fibrin. The bioink can be supplemented with solid material. Using LAB, we stacked 15 layers of nano-sized hydroxyapatite [61]. As observed by scanning electron microscopy, the material was highly compacted after the printing. In order to build a 3D porous structure, some sacrificial material layer [62] could be printed concomitantly to produce the desired porosity into the material. This experiment shows that it is possible to use the high-throughput capability of LAB to concentrate and stack nano-sized particles from an aqueous solution. However, LAB is not completely suited for building tissue constructs with cm^3 size and handling capacity. Indeed, the characteristic droplet volume is on the order of 1 pl [24].

6.4.2 **Biopapers**

Some material can be used in the layer-by-layer approach to provide volume and/ or biochemical properties that the liquid bioink may not supply once it is patterned to stabilize the pattern of the printed cells and support the construct in its whole. Taking into account the diffusion limit of oxygen, nutrients, and waste removal, thin, solid films of biomaterials, called biopapers, are used as stackable layers that support bioprinted material or cells. The behavior of such three-dimensional constructs was addressed using electrospun 100-μm-thick scaffolds of polycaprolacton (PCL) and cells printed by LAB (Figure 6.12 and [24]). Also, PLGA/hydrogel (type I collagen or Matrigel^{TM}) biopapers have been used to print HUVEC [63]. Such biopapers may be patterned to provide additional control on possible cell migration and differentiation. Alternatively, hydrogel could be patterned itself (see Figure 6.10) prior to receiving bioprinted cells, or cell co-culture can be designed to physiologically stabilize one another in the initial printed pattern [31,45].

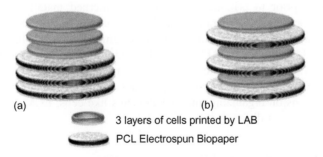

(a)   (b)

🔵 3 layers of cells printed by LAB

⬜ PCL Electrospun Biopaper

FIGURE 6.12

Simulated conventional seeding of stacked PCL biopapers with MG63 cells printed by LAB (a), compared to layer-by-layer organization using LAB of MG63 cells and PCL electrospun biopapers (b). The same numbers of cells and of PCL scaffolds have been used in both conditions. PCL: polycaprolacton.

6.5 Applications

In addition to liquids and all kinds of organic molecules in solution, numerous studies have shown successful laser-assisted printing of a fairly broad range of prokaryotic cells and eukaryotic cells (see [23] for a review). Considering human primary cells in particular, human umbilical vein endothelial cells (HUVEC) and human umbilical vein smooth muscle cells (HUVSMC) [31,45], human mesenchymal stem cells [45,51,59], adipose-tissue derived stem cells (ADSC) and endothelial colony-forming cells (ECFC) [64], and human bone-marrow derived osteoprogenitors (HOP) [46] have all been printed by LAB.

6.5.1 Multicolor printing at the microscale

Since biological tissues are composed of multiple components in close interactions with one another (cells of different types, proteins, and other components of the ECM), not only 3D structures but also "multicolor" printing—that is, the printing of multiple biological components—should be considered in tissue engineering products. Together with the high-resolution printing capability of LAB, it is possible to print different cell types in close contact with one another, with a high cell concentration, according to a desired spatial organization [27]. The micrometric printing resolution achievable by LAB for multiple cell types and materials is consistent with the study of cell-to-cell or cell-to-material interactions as well.

6.5.2 LAB engineered stem cell niche and tissue chips

Printing bioactive factors (e.g., morphogens, growth factors) onto inserted materials to regulate cell fate (in terms of migration or differentiation, for example) or to position cells at the desired coordinates is achievable by LAB. Another, yet nonexclusive, approach would be to use LAB for printing resolutive patterns of material itself prior to printing cells onto the patterned material [27]. Studies would also deal with generating artificial cell niches by codepositing a suitable combination of stem cells with ECM components [65]. Indeed, embryonic stem cells printed by MAPLE-DW—formed embryoid bodies with retained pluripotency [53]. In relation to these issues, mechanical and topological cues should be studied using bottom-up approaches for engineering tissues. Future studies will focus on organizing multiple elements like cells, ECM-like materials, and growth factors at different scales of histology.

6.5.3 Addressing the perfusion issue in tissue engineering

The main limitation with respect to thick cellularized tissue constructs is the time required for the assembly and maturation of a perfused vascular network throughout the entire tissue construct. In certain cases, the assembly and maturation time might be longer than the cell survival time. Micropattern-guided vasculogenesis might help quicken vascular lumen formation as well as branching between the host and the tissue construct. To this aim, endothelial cords have auto-assembled consecutively to endothelial cell alignment with cell-scale accuracy by LAB [45]. Such an approach to modelized endothelial tube formation might be fruitful in the field of vasculogenesis- and angiogenesis-related research.

6.5.4 Laser-assisted engineering of transplants

To our knowledge, two studies report in vivo transplantation of LAB engineered tissue constructs. Gaebel et al. used LAB to fabricate a cardiac patch for cardiac regeneration in a rat model of acute myocardial infarction [31]. LAB was used to pattern a co-culture of HUVEC and HMSC onto a 300-μm-thick, 8-mm-diameter disk of polyester urethane urea. The healing potential of the patterned patch compared similarly to the unpatterned patch, demonstrating the suitability of the LAB procedure for tissue regeneration and suggesting that patterning may favor faster vasculogenesis and grafting.

The influence of the three-dimensional organization of MG63 cells and electrospun 100-μm-thick PCL biopapers was evaluated regarding cell proliferation in vitro and in vivo [46]. For this purpose, a layer-by-layer sandwich model of assembly was compared to a control hybrid material made of the same amount of material with an alternative 3D arrangement. These constructs were both evaluated in vitro and in vivo in a mouse calvarial defect reconstruction model. Results underscore the benefit of the layer-by-layer approach to encapsulate cells within a sandwich of PCL biopapers, either in vitro or in vivo, as far as cell viability and cell proliferation are concerned.

In both reports, cell patterning at cell-level resolution was not studied. Increasing the resolution may help guide faster tissue specific organization like faster vasculogenesis, which may support faster morphogenesis or healing process.

6.5.5 In vivo printing

To the best of our knowledge, all previous studies report on LAB fabrication in vitro. Our team has performed some preliminary assays for in vivo printing (Figure 6.13 and [66]). More precisely, the purpose of our study was to fill a critical-sized calvarial bone defect in mouse by printing hydroxyapatite. A specific mouse holder was designed to position the surface of mice dura mera instead of the quartz substrate. Then, 30 layers of hydroxyapatite were printed inside one defect. The histological results have shown that the material was present in the test defects of all groups. However, bone repair was inconstant. As a conclusion, we showed that in vivo bioprinting is possible. Future experiments in this model should improve the mechanical and biological properties of the printed material, as well as cell printing resolution.

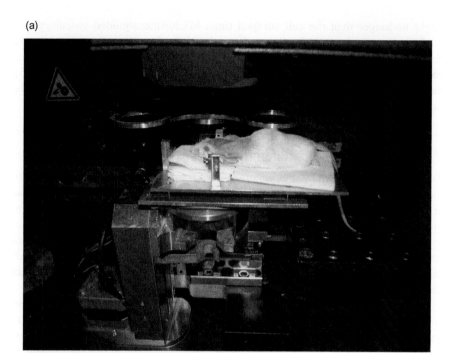

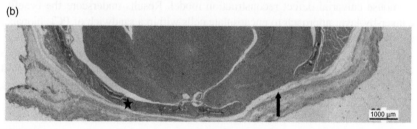

FIGURE 6.13

In vivo printing reconstruction of calvarial bone defect in a mouse. The calvarial defect is placed under the ribbon (a). Histological section of mouse calvarial defect (b). Complete bone repair on the test side (star) was observed in one sample after 3 months. The bone defect control site remains unrepaired (arrow).

CONCLUSION

In this chapter, we show that LAB of cells requires an understanding of the dynamics of vapor bubbles that govern the droplet ejection. Consequently, the bioink should be designed accordingly, and spatio-temporal proximity of consecutive laser-induced jets should be considered for optimal printing resolution. Several studies taken together demonstrate the capability of LAB to print virtually all cell types, although many human cell types remain to be validated. These cells can be printed onto numerous biomaterials, either solid or gel, like biopolymers and nano-sized particles of hydroxyapatite. The potential of the LAB to fabricate functional cells containing transplants for tissue repair has been demonstrated, together with the possibility to shunt the transplantation process by operating LAB directly in vivo.

Combining LAB with other laser-assisted processes, such as machining and polymerization, should be addressed with specific attention on integrating these different processes in the same workstation to guarantee subcellular resolution. However, because LAB manages droplets of a volume on the order of pL, this technology has not overcome the typical trade-off between resolution and manufacturing scale. As a direct write method of living cells, LAB can be combined with other tissue engineering methods [58].

Concerning the layer-by-layer microfabrication of functional tissues that mimic their in vivo counterparts, it remains to be determined whether the exact reproduction of the histoarchitecture of living tissue is crucial—in other words, to what extent and resolution cellular self-assembly has to be guided. How multiple cell types (e.g., endothelial cells, biliary epithelial cells, and hepatocytes) might polarize and self-assemble to form functional units (e.g., a liver lobule) could be addressed by miniaturized TE approaches. Future studies involving pattern formation in morphogenesis—specifically the relationship between form and function— should address this aim. Moreover, the engineering of realistic tissue constructs will help further the understanding of tissue physiology and function; this, in turn, will refine TE strategies and optimize blueprints.

Acknowledgments

We acknowledge financial support from GIS-AMA (Advanced Materials in Aquitaine), ANR (Agence Nationale pour la Recherche), and Région Aquitaine.

References

[1] Lalan S, Pomerantseva I, Vacanti JP. Tissue engineering and its potential impact on surgery. World J Surg 2001;25(11):1458–66.

[2] Lysaght MJ, Jaklenec A, Deweerd E. Great expectations: private sector activity in tissue engineering, regenerative medicine, and stem cell therapeutics. Tissue Eng Part A 2008;14(2):305–15.

[3] Yamada KM, Cukierman E. Modeling tissue morphogenesis and cancer in 3D. Cell 2007;130(4):601–10.

[4] Inamdar NK, Borenstein JT. Microfluidic cell culture models for tissue engineering. Curr Opin Biotechnol 2011;22(5):681–9.

[5] Hutmacher DW. Biomaterials offer cancer research the third dimension. Nat Mater 9 (2):90–3.

[6] Baquey, C, Dupuy B. 1989. Hybrid artificial organs: concepts and development. Les Editions INSERM.

[7] Langer R, Vacanti J. Tissue engineering. Science 1993;260(5110):920–6.

[8] McGuigan AP, Sefton MV. Vascularized organoid engineered by modular assembly enables blood perfusion. Proc Natl Acad Sci USA 2006;103(31):11461−6.

[9] McGuigan AP, et al. Cell encapsulation in sub-mm sized gel modules using replica molding. PLoS ONE 2008;3(5):e2258.

[10] Mironov V, et al. Organ printing: Tissue spheroids as building blocks. Biomaterials 2009;30(12):2164−74.

[11] Hutmacher DW. Scaffolds in tissue engineering bone and cartilage. Biomaterials 2000;21(24):2529−43.

[12] Mironov V. Organ printing: computer-aided jet-based 3D tissue engineering. Trends Biotechnol 2003;21(4):157−61.

[13] Jakab K, et al. Tissue engineering by self-assembly and bio-printing of living cells. Biofabrication 2010;2(2):022001.

[14] Guillemot F, Mironov V, Nakamura M. Bioprinting is coming of age: report from the international conference on bioprinting and biofabrication in Bordeaux (3B'09). Biofabrication 2010;2(1):010201.

[15] Klebe RJ, et al. Cytoscription: computer controlled micropositioning of cell adhesion proteins and cells. Methods Cell Sci 1994;16(3):189−92.

[16] Klebe RJ. Cytoscribing: A method for micropositioning cells and the construction of two- and three-dimensional synthetic tissues. Exp Cell Res 1988;179(2):362−73.

[17] Boland T, et al. Application of inkjet printing to tissue engineering. Biotechnol J 2006;1(9):910−7.

[18] Nakamura M, et al. Biocompatible inkjet printing technique for designed seeding of individual living cells. Tissue Eng 2005;11(11−12):1658−66.

[19] Saunders RE, Gough JE, Derby B. Delivery of human fibroblast cells by piezoelectric drop-on-demand inkjet printing. Biomaterials 2008;29(2):193−203.

[20] Brisbane, 1971. PATTERN DEPOSIT BY LASER. US patent by Brisbane on Feb 2, 1971. Patnet # 3,560,258 available on http://www.google.com/patents?hl = en&lr = &vid = USPAT3560258&id = x4FUAAAAEBAJ&oi = fnd&dq = Pattern + deposit + by + laser + filed + for + patent +1967 + %22BRISBANE%22&printsec = abstract#v = onepage&q = Pattern%20deposit%20by%20laser%20filed%20for%20 patent%201967%20%22BRISBANE%22&f = false.

[21] Young D, et al. Plume and jetting regimes in a laser based forward transfer process as observed by time-resolved optical microscopy. Appl Surf Sci 2002;181−7, 197−8.

[22] Bohandy J, Kim BF, Adrian FJ. Metal deposition from a supported metal film using an excimer laser. J Appl Phys 1986;60(4):1538.

[23] Schiele NR, et al. Laser-based direct-write techniques for cell printing. Biofabrication 2010;2(3):032001.

[24] Guillemot F, Souquet A, et al. High-throughput laser printing of cells and biomaterials for tissue engineering. Acta Biomater 2010;6(7):2494−500.

[25] Barron, Jason A, Krizman DB, Ringeisen, Bradley R. Laser printing of single cells: statistical analysis, cell viability, and stress. Ann Biomed Eng 2005;33(2):121−30.

[26] Barron JA, et al. Biological laser printing: a novel technique for creating heterogeneous 3-dimensional cell patterns. Biomed Microdevices 2004;6(2):139−47.

[27] Guillotin Bertrand, et al. Laser assisted bioprinting of engineered tissue with high cell density and microscale organization. Biomaterials 2010;31(28):7250−6.

[28] Duncan AC, et al. Laser microfabricated model surfaces for controlled cell growth. Biosens Bioelectron 2002;17(5):413−26.

[29] Claeyssens F, et al. Three-dimensional biodegradable structures fabricated by two-photon polymerization. Langmuir 2009;25:3219−23.

[30] Lazare S, et al. Surface foaming of collagen, chitosan and other biopolymer films by KrF excimer laser ablation in the photomechanical regime. Appl Phys A-Mater 2005;81(3):465−70.

[31] Gaebel R, et al. Patterning human stem cells and endothelial cells with laser printing for cardiac regeneration. Biomaterials 2011;32(35):9218−30.

[32] Guillemot F, Souquet A, et al. Laser-assisted cell printing: principle, physical parameters versus cell fate and perspectives in tissue engineering. Nanomedicine 2010;5 (3):507−15.

[33] Duocastella M, et al. Jet formation in the laser forward transfer of liquids. Appl Phys A 2008;93(2):453−6.

[34] Duocastella M, et al. Study of the laser-induced forward transfer of liquids for laser bioprinting. Appl Surf Sci 2007;253(19):7855−9.

[35] Mezel C, et al. Self-consistent modeling of jet formation process in the nanosecond laser pulse regime. Phys Plasmas 2009;16(12):123112−12.

[36] Prosperetti A. A generalization of the Rayleigh−Plesset equation of bubble dynamics. Phys Fluids 1982;25(3):409−10.

[37] Xiu-Mei L, et al. Growth and collapse of laser-induced bubbles in glycerol−water mixtures. Chin Phys B 2008;17(7):2574−9.

[38] Pearson A, et al. Bubble interactions near a free surface. Eng Anal Bound Elem 2004;28(4):295−313.

[39] Robinson PB, et al. Interaction of cavitation bubbles with a free surface. J Appl Phys 2001;89(12):8225−37.

[40] Duocastella M, et al. Time-resolved imaging of the laser forward transfer of liquids. J Appl Phys 2009;106(8):084907.

[41] Brown MS, Kattamis NT, Arnold CB. Time-resolved dynamics of laser-induced micro-jets from thin liquid films. Microfluid Nanofluid 2011;11(2):199−207.

[42] Brown MS, Kattamis NT, Arnold CB. Time-resolved study of polyimide absorption layers for blister-actuated laser-induced forward transfer. J Appl Phys 2010;107 (8):083103−083103−8.

[43] Engler AJ, et al. Matrix elasticity directs stem cell lineage specification. Cell 2006;126(4):677−89.

[44] Engler AJ, et al. Multiscale modeling of form and function. Science 2009;324 (5924):208−12.

[45] Wu PK, Ringeisen BR. Development of human umbilical vein endothelial cell (HUVEC) and human umbilical vein smooth muscle cell (HUVSMC) branch/stem structures on hydrogel layers via biological laser printing (BioLP). Biofabrication 2010;2(1):014111.

[46] Catros S, Guillotin B, et al. Effect of laser energy, substrate film thickness and bioink viscosity on viability of endothelial cells printed by laser-assisted bioprinting. Appl Surf Sci 2011;257(12):5142−7.

[47] Othon CM, et al. Single-cell printing to form three-dimensional lines of olfactory ensheathing cells. Biomed Mater 2008;3(3):034101.

[48] Ringeisen, Bradley R, et al. Laser printing of pluripotent embryonal carcinoma cells. Tissue Eng 2004;10(3−4):483−91.

[49] Wang Wei, et al. Study of impact-induced mechanical effects in cell direct writing using smooth particle hydrodynamic method. J Manuf Sci Eng 2008;130(2): 021012−10.

[50] Catros S, Fricain J-C, et al. Laser-assisted bioprinting for creating on-demand patterns of human osteoprogenitor cells and nano-hydroxyapatite. Biofabrication 2011;3 (2):025001.

[51] Gruene M, et al. Laser printing of stem cells for biofabrication of scaffold-free autologous grafts. Tissue Eng Part C: Methods 2010:00830145320029.

[52] Hopp B, et al. Survival and proliferative ability of various living cell types after laser-induced forward transfer. Tissue Eng 2005;11(11−12):1817−23.

[53] Raof NA, et al. The maintenance of pluripotency following laser direct-write of mouse embryonic stem cells. Biomaterials 2011;32(7):1802−8.

[54] Colina M, et al. DNA deposition through laser induced forward transfer. Biosens Bioelectron 2005;20(8):1638−42.

[55] Serra P, et al. Liquids microprinting through laser-induced forward transfer. Appl Surf Sci 2009;255.

[56] Dinca V, et al. Directed three-dimensional patterning of self-assembled peptide fibrils. Nano Lett 2008;8(2):538−43.

[57] Schiele NR, et al. Laser direct writing of combinatorial libraries of idealized cellular constructs: biomedical applications. Appl Surf Sci 2009;255(10):5444−7.

[58] Guillotin B, Guillemot F. Cell patterning technologies for organotypic tissue fabrication. Trends Biotechnol 2011;29(4):183−90.

[59] Koch L, et al. Laser printing of skin cells and human stem cells. Tissue Eng Part C: Methods 2009;: 091221133515000.

[60] Gruene M, et al. Adipogenic differentiation of laser-printed 3D tissue grafts consisting of human adipose-derived stem cells. Biofabrication 2011;3:015005.

[61] Catros S, Guillemot F, et al. Layer-by-layer tissue microfabrication supports cell proliferation in vitro and in vivo. Tissue Eng Part C: Methods 2011;111107133249008.

[62] Lee W, et al. Multi-layered culture of human skin fibroblasts and keratinocytes through three-dimensional freeform fabrication. Biomaterials 2009;30(8):1587−95.

[63] Pirlo RK, et al. PLGA/hydrogel biopapers as a stackable substrate for printing HUVEC networks via BioLP^{TM}. Biotechnol Bioeng 2012;109(1):262−73.

[64] Gruene Martin, et al. Laser printing of three-dimensional multicellular arrays for studies of cell−cell and cell−environment interactions. Tissue Eng Part C: Methods 2011;110629135038006.

[65] Lutolf MP, Blau HM. Artificial stem cell niches. Adv Mater 2009;21 (32−33):3255−68.

[66] Keriquel V, et al. In vivo bioprinting for computer- and robotic-assisted medical intervention: preliminary study in mice. Biofabrication 2010;2(1):014101.

[67] Longuet-Higgins MS. Bubbles, breaking waves and hyperbolic jets at a free surface. J Fluid Mech Digit Arch 1983;127(1):103−21.

Further Reading

Blake JR, Gibson DC. Growth and collapse of a vapour cavity near a free surface. J Fluid Mech Digit Arch 1981;111(1):123−140.

Brown MS, Kattamis NT, Arnold CB. Time-resolved study of polyimide absorption layers for blister-actuated laser-induced forward transfer. J Appl Phys 2010;107(8):083103.

M. Duocastella, et al., Novel laser printing technique for miniaturized biosensors preparation. Sens Actuators, B 2010;145(1):596−600.

Fabien Guillemot, et al., Laser-Assisted bioprinting to deal with tissue complexity in regenerative medicine. Mrs Bull 2011;36(12):1015−9.

The Modular Approach

7

Ema C. Ciucurel, M. Dean Chamberlain, Michael V. Sefton

Department of Chemical Engineering and Applied Chemistry, Institute of Biomaterials and Biomedical Engineering, University of Toronto, Toronto, Canada

CONTENTS

INTRODUCTION

Modular tissue engineering was first introduced by the Sefton laboratory as a means of building intrinsically vascularized tissue constructs [1,2]. Instead of using the conventional method of seeding a "large" porous scaffold with cells, the group proposed seeding smaller constructs, which they called "modules," and packing these modules together to obtain a larger tissue construct. With this method of assembly, empty channels form among the modules upon packing of several modules, and these channels are interconnected to provide the porosity that is otherwise generated by starting with a conventional scaffold. Since these modules were covered with endothelial cells (EC) prior to assembly, the channels were lined with EC and amenable to perfusion (with blood), similarly to a vascular network—hence the intrinsically vascularized nature of the modular approach. Vascular supporting cells (such as smooth muscle cells, SMC [3], or mesenchymal stromal cells, MSC [4]) or therapeutic cells of interest (cardiomyocytes [5], islets [6], or others) can be embedded inside the modules. Modules containing different supporting or therapeutic cells can also be mixed together in a desirable ratio to build more complex functional tissue structures (Figure 7.1).

The modular approach, when first proposed by the Sefton group, aimed to address the lack of an internal vascular network, a key issue in tissue engineering. A vascular supply is required to supply nutrients to the cells, to remove waste products, for gas exchange, and for circulation of signaling molecules. Due to diffusion limitations within tissues, all cells must be within 100 to 200 μm of a blood vessel [7]. Therefore, from a tissue engineering perspective, an immediate vascular supply is required to preserve cell viability and tissue function at clinically relevant sizes. Compared to the conventional approach of seeding cells on top of scaffolds and relying mostly on host cell infiltration to remodel and

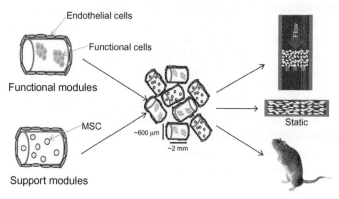

FIGURE 7.1

Modular tissue engineering with endothelialized building blocks. Different modular units that contain different cell types can be made separately and then mixed together in various ratios to form more complex structures. These complex structures can then be used for in vitro modeling of tissues (under static or flow conditions) or implanted. There are two basic types of modular units. Functional modules contain the therapeutic cells or the cells of interest for study. Support modules contain cell types (such as mesenchymal stromal cells) that support the function of the therapeutic cells or improve the vascularity of the implanted construct.

vascularize the scaffold (slow process), the modular approach has the advantage of a "built-in" vascular network, with vascular channels preformed by design between the modules. The small size of the modules (~2 mm long and 0.6 mm in diameter starting size, which is typically further contracted by embedded functional cells and EC) also ensures that the cells embedded in any individual module will not experience hypoxia [8]. Hypoxia is typical in thick tissue constructs fabricated using the conventional approach. Other advantages of modular tissue engineering are that the modular design allows for uniform cell seeding within the construct (by controlling the cell density within each of the individual modular building blocks), as well as controlled mixing of different cell populations (by mixing together modules encapsulating different functional cells, in different ratios), and this approach is scalable (increasingly larger tissue constructs can be made by assembling increasing numbers of modular building blocks). The modules can also be easing their utility in some situations. Moreover, from a biomimetic standpoint, the modular approach recapitulates the "design" of native tissues and organs, which are often composed of repeating functional units. For example, pancreatic islets, hepatic lobules, muscle fibers, and so on each consist of repeating functional units to form larger structures. Many research groups have expanded the use of modular tissue engineering from what was initially envisioned as a method of building intrinsically vascularized and scalable constructs to a method of controlling tissue architecture and building larger tissue constructs from the "bottom up" [9−14], thereby expanding the definition of a "module" (Figure 7.2).

This chapter focuses on the different variations of the modular tissue engineering approach that are currently being explored. Figure 7.2 illustrates some of these current trends. Modules containing encapsulated cells and fabricated using different natural or synthetic biomaterials are used to either create intrinsically vascularized tissue constructs or as building blocks for generating larger tissue constructs in vitro, sometimes through controlled assembly of the modules, with the goal of controlling the architecture of the final assembled construct. Modules integrated in microfluidics systems are being developed as in vitro tissue models for studying cell interactions, for drug testing, and for performing different bioassays in a controlled, three-dimensional, perfusable cell culture environment, with multiple cell types co-cultured inside the device.

7.1 Materials used to fabricate the modular building blocks

Both natural extracellular matrix (ECM) components and synthetic polymers are being used to make modular tissue engineered constructs. Collagen (an ECM component) and poly(ethylene glycol) (PEG)-based synthetic polymers are the most widely used.

(a) Types of modular units

Endothelialized

Advantages

Integrated vascular component

Can localize cells within each unit

Disadvantages

No control of assembly

Low cell density

Defined shape

Advantages

Control of assembly

Easy to control location of cells

Disadvantages

Needs UV crosslinkers (toxic)

To maintain shape uses nondegradable matrix

Cell sheets

Advantages

Covers large areas

Can layer cell types

Disadvantages

Not space filling

Hard to make thick tissues

Spheroids

Advantages

High cell density

Good cell viability

Disadvantages

Small size

Can have necrotic cores

Microcarriers

Advantages

High cell density

Good cell attachment

Disadvantages

Small size

Lack of tissue architecture

Tissue slice

Advantages

Tissue architecture

Contains all cell types

Disadvantages

In vitro only

Minimal thickness

(b) Types of materials used for modular tissue engineering

Natural matrix

Advantages

High cell attachment

Good cell viability

Disadvantages

Does not hold its shape

Deforms under flow

Synthetic matrix

Advantages

Holds its shape

Tunable stiffness

Disadvantages

Poor cell attachment

Poor cell viability

FIGURE 7.2

Approaches to modular tissue engineering. (a) Types of modular units (schematic cross section) that are currently being used in tissue engineering. Advantages and disadvantages of each method are listed. (b) The classes of materials that are being used for scaffolds in modular tissue engineering with advantages and disadvantages.

(a) Automated cutter method of making modules

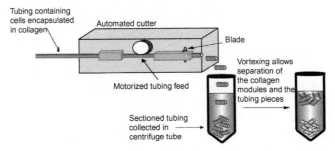

(b) Air plug method of making modules

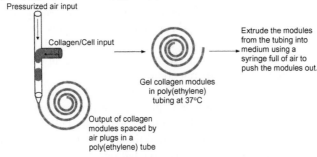

FIGURE 7.3

Methods of making endothelialized modules. (a) Automated cutter method. Collagen with or without embedded cells is gelled in a length of poly(ethylene) tubing. The tubing is then cut into small (~2 mm in length) segments and collected in a 50-mL tube containing medium. After 1 hour of incubation at 37°C, the tube is vortexed, separating the tubing and the modules. The modules are collected and coated with EC. (b) Air plug method. Using a T junction, droplets of the collagen with or without embedded cells are generated and loaded in the poly(ethylene) tubing. The collagen is gelled at 37°C, and the modules are then forced out of the tubing using a syringe, and collected in a tube that contains medium. The modules are then coated with EC.

7.1.1 **ECM-based materials**

The Sefton group fabricated cylindrical collagen modules by gelling a neutralized collagen type I solution (with or without embedded functional cells) at 37°C inside a polyethylene tubing and then cutting the tubing into small pieces using an automatic custom-made tube cutter (Figure 7.3A). The resulting cylindrical pieces of collagen were separated from the tubing by vortexing and were coated with endothelial cells through a combination of static and dynamic seeding. These endothelialized modules were then randomly packed together to obtain larger, intrinsically vascularized modular tissue constructs [1,2,15]. In a newer fabrication method, the collagen liquid was sheared into liquid modules upon contact with a perpendicular stream of pressurized air, resulting in individual collagen modules separated by air spaces inside the polyethylene tubing. After gelation of the collagen inside the tubing, the modules were simply expelled using a syringe, thus eliminating the need for cutting the tubing, which is time consuming and therefore less practical for larger batches [16] (Figure 7.3B). A combination of ECM materials was also used, with fibronectin coating of collagen modules shown to improve the survival of EC upon subcutaneous implantation in a SCID/Bg animal model [17].

Other groups use collagen to fabricate modules. Using a soft-lithography method, the Whitesides group fabricated collagen microgels inside a polydimethylsiloxane (PDMS) mold. The microgels had various shapes (circular, square, cross), depending on the shape of the mold, and contained encapsulated NIH/3T3 fibroblasts, HepG2 liver cells, or primary rat cardiomyocytes. Matrigel™ and agarose modules were also fabricated using the same method [18]. The Demirci group used a droplet-based system, with SMC suspended in collagen and deposited in droplets under precise spatial control [19].

Gelatin, a denatured form of collagen, has also been used to build modular constructs, either in the form of gelatin microcarriers or as photocrosslinkable gelatin derivatives. In one study, gelatin microcarriers seeded with fibroblasts were assembled into a three-dimensional dermal tissue [20]. In a different study, human amniotic mesenchymal stem cells were seeded onto gelatin microcarriers, underwent osteogenic differentiation, and were assembled into a bone construct in a perfusion chamber [21]. Gelatin hydrogels were fabricated by synthesizing a gelatin methacrylate derivative and exposing the synthesized polymer solution to UV light (photocrosslinking) to form a gel [22]. Similarly, gelatin methacrylate mechanically reinforced with carbon nanotubes, or mixed with pullulan methacrylate (a fungal polysaccharide), has been photocrosslinked into microgel building blocks [23,24].

Another natural material, hyaluronic acid, has also been used to fabricate modules. Cells (NIH/3T3 fibroblasts or murine embryonic stem cells R1 strain) were mixed with a methacrylate derivative of hyaluronic acid (MeHA) and deposited in a PDMS mold [25]. After exposure to UV light and gelling of the polymer solution, hydrogel modules of different shapes and sizes (as defined by the micropatterns of the PDMS mold) were collected from the mold. Although not done here, these hydrogels could presumably be assembled together to build a larger modular tissue. However, as with all systems that rely on photocrosslinking for formation of the gels, both the exposure time to UV light and the concentration of the photoinitiator required to initiate photocrosslinking need to be minimized to preserve cell viability. This is often not a trivial requirement, since short UV exposure times and low photoinitiator concentrations also make gelling of the hydrogels difficult, leading to soft hydrogels that are difficult to manipulate, let alone assemble into a larger porous structure without collapsing.

7.1.2 **PEG-based materials**

Modular tissue engineering is also amenable to the use of purely synthetic materials that can form hydrogels. PEG is a synthetic polymer commonly used in tissue engineering to make hydrogel scaffolds. Specifically in modular tissue engineering, PEG has been used extensively to fabricate hydrogel building blocks of different sizes and shapes (circles, squares, lock-and-key structures, saw-shaped microgels, etc.) that can then be assembled into larger structures with controlled architecture [26−29]. PEG-based hydrogels are typically obtained through placement under a photomask of a solution of PEG polymer functionalized with methacrylate or acrylate groups and exposure to UV light for photocrosslinking. By varying the size and shape of the photomask and the thickness of the PEG layer, hydrogels of various sizes and shapes are obtained. These hydrogel building blocks are then assembled in a random or controlled manner using various techniques, such as acoustic assembly and thermodynamically driven assembly in multiphase liquid systems (see the section on assembly with microfluidic approaches) [26,27,30]. The assembled final structures can be further stabilized through a secondary UV crosslinking step [26,27].

PEG-based hydrogel systems offer tremendous opportunities in terms of controlling the initial architecture of the building blocks. However, including cells in these PEG-based gels can be difficult due to the toxicity of the photoinitiator at the concentration and UV exposure time required to form mechanically strong gels through photocrosslinking. Unlike PEG, ECM materials such as collagen do not require such chemical crosslinking to form gels; the process of making collagen gels is entirely cell compatible, as collagen gels are formed by simply incubating a neutralized collagen solution at 37°C, the standard cell culture temperature. Moreover, PEG-based systems generally lack the biological properties required to build functional tissues (at least without further modification), starting with their low cell binding and nondegradable properties. They are therefore generally limited to proof-of-concept studies, with a focus on controlling the initial architecture but not the long-term functionality of the assembled tissue construct; most of these studies only investigated the short-term (i.e., typically on the order of a few hours to a few days) survival of cells. Mixing the PEG-based polymer solution with cell-compatible ECM components, such as collagen, can be useful for maintaining cell viability in these photocrosslinkable hydrogels. There are, of course, many reports in the tissue engineering and biomaterials literature (although not necessarily in the context of modular tissue engineering) describing PEG-based scaffolds that have been chemically or physically modified to include cell-responsive components (cell attachment, enzymatic degradation, etc.). For example, in one study reporting the fabrication of 3D hepatic tissues from building blocks, primary hepatocytes were successfully cultured on PEG-based hydrogels functionalized with RGD sequences and assembled into a 3D construct. The hepatocytes maintained liver-specific functions over 12 days in culture in vitro under perfusion [31].

Poloxamine (a four-arm block co-polymer of poly(propylene oxide) and poly(ethylene oxide) with properties similar to PEG) has also been used in modular tissue engineering. Methacrylate groups were chemically added to poloxamine, and the synthesized polymer was gelled inside a polyethylene tubing by exposure to UV light [32,33]. To improve cell attachment and viability of cells embedded within or seeded onto poloxamine modules, poloxamine was mixed with collagen [32,33], chemically modified to introduce positively charged groups into the scaffold biomaterial [34,35], chemically linked to polylysine peptide chains [36], or poloxamine modules were coated with laminin, another ECM protein [37]. Since the ability of the cells to remodel the tissue engineered construct is also an important design factor, remodelable poloxamine-based modules were also fabricated by synthesizing a lactoyl-poloxamine methacrylate derivative [38]. In a related approach, PEG methacrylate was mixed with gelatin methacrylate and photo crosslinked together to obtain microgels that combined the favorable mechanical properties of PEG and the cell attachment and degradable properties of gelatin [39].

7.2 Intrinsically vascularized tissue constructs

When first presented by McGuigan and Sefton, modular tissue engineering was envisioned as a method to create intrinsically vascularized tissue constructs, with a quiescent layer of EC lining the module surface and creating nonthrombogenic vessel-like channels between the modules. Modules coated with human umbilical vein endothelial cells (HUVEC) were indeed nonthrombogenic in vitro, which is an important requirement for implantable vascular structures [40,41]. When exposed to whole blood, the endothelialized modules showed reduced platelet activation (reduced microparticle formation) and platelet-leukocyte associations (platelet positive events in a leukocyte gated sample by flow cytometry) compared with the non-EC-coated collagen modules [40]. The EC-coated modules also delayed clotting when exposed to whole blood on a rocking platform, and there was limited fibrin and platelet deposition on the surface of the modules under static conditions [40]. The endothelialized constructs also enabled whole blood perfusion, with limited platelet loss/deposition in the case of the endothelialized modules [40]. HUVEC on collagen modules also expressed low levels of tissue factor (TF) and high levels of thrombomodulin (TM), suggestive again of their nonthrombogenic phenotype [41]. While nonthrombogenicity was confirmed in vitro, in vivo studies showed that endothelial cells did not remain attached to the collagen gel and in a quiescent state. Rather, they drove a remodeling process that resulted in a chimeric vasculature that was different from what was originally envisaged.

7.2.1 In vivo fate of endothelialized modules

In vivo, host response and remodeling of the modular construct become decisive factors for the outcome of the implant. Like all tissue constructs, biofabrication of implantable modular tissues should include in vivo remodeling as a fundamental design parameter. The number of in vivo reports using the modular tissue engineering approach is, however, still limited, at least by comparison to the larger number of in vitro studies.

In one of the first in vivo modular tissue engineering studies by Gupta et al. [42], collagen modules (without embedded cells) coated with HUVEC were implanted in the omental pouch of nude rats, with temporary depletion of macrophages (using clodronate liposome injections) to improve EC survival. Histology images showed that the modules randomly assembled and the channels formed between the modules persisted for at least 14 days. However, the HUVEC did not remain on the surface of the modules, as initially predicted, but instead migrated off the modules and within 7 days after implantation started to form primitive blood vessels in the area between the modules. There were signs of incipient vascularization of the tissue, with host rat EC, SMC, and red blood cells investing these primitive blood vessels over the 14-day course of the experiment. However, the EC survival and blood vessel formation were still limited, presumably due to the inflammatory and immune host response directed against the implanted human cells in this only partially immune compromised animal model; clodronate had only a temporary depleting effect on invading macrophages.

Overall, this first in vivo modular tissue engineering study brought attention to the importance of properly modulating the remodeling and host response if the end goal is to obtain a vascularized tissue construct. More attention to modulating the remodeling process in vivo is required and perhaps less on the initial shape and architecture of the constructs prefabricated in vitro, which will most likely eventually be remodeled in vivo. For example, the degree of confluence or quiescence of the attached EC appears to be less important than it was originally deemed. Creating a construct that will drive a favorable remodeling response in vivo is perhaps the key, at least for applications where the end goal is not an in vitro tissue model.

In a subsequent study, collagen modules coated with rat aortic EC (RAEC) were implanted in the omental pouch of immunosuppressed (allogeneic) Sprague-Dawley rats [43]. No T cells (TCR $\alpha\beta+$) were observed in the implant area, and the inflammatory cell numbers (CD68+) were reduced in the drug immunosuppressed animals compared with the untreated control animals. Similarly to the previous study, the authors reported the remodeling of the modular implants. Initially (day 3 after the implant), the RAEC were still seen on the surface of the modules. At a later time (day 7), the RAEC had migrated off the surface of the modules and had started forming primitive blood vessels. At even later time points (days 14, 21, and 60), the blood vessels became invested with SMC and erythrocytes (suggestive of connection to the host vasculature). Both host- and donor-derived EC formed the endothelial cell lining of these blood vessels. Figure 7.4A shows a representative picture of the implant 21 days after surgery.

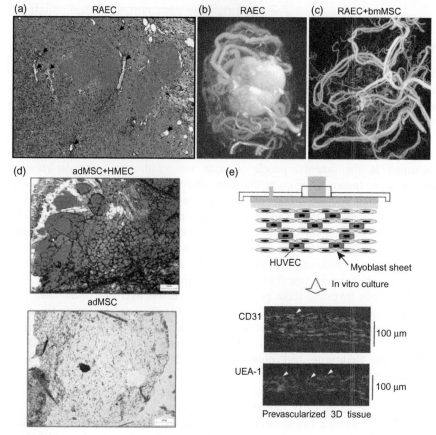

FIGURE 7.4

In vivo modular tissue engineering. (a) Trichrome image of rat aortic EC (RAEC) coated modules (without embedded cells) implanted into a rat omental pouch for 21 days. Blood vessels (arrows) formed around and near the modules (outlined by dashed red line). (b) MicroCT image of the vascular system formed from the RAEC modules (of [a]) implanted into a rat omental pouch after 21 days. There was a large, leaky core on the microCT images, which is evidence that these vessels were immature and leaky. (c) MicroCT image of the vascular system formed from RAEC-coated bmMSC embedded modules implanted into a rat omental pouch after 21 days. The addition of the MSC decreased the size of the leaky core, consistent with the improved maturation of the blood vessels. (d) Fat development (Oil red O staining) within modules containing human adipose-derived MSC (adMSC) with or without human microvascular EC (HMEC) in a SCID/Bg mouse after 90 days (subcutaneous injection). Without the presence of the HMEC, the adMSC did not develop into a fat pad (bottom panel), suggesting that the vascularization is a prerequisite for the formation of the fat pad. (e) The fabrication of a cell sheet construct containing myoblasts and HUVEC (top panel). Bottom panels: Cross-sectional images of the construct showing vascularization (white arrowheads). The sheet is stained with antihuman CD31 antibody (green) or UEA-I (red) and counterstained with Hoechst 33342 (blue).

MicroComputed Tomography (microCT) studies performed on the whole omental pouch at 21 days (Figure 7.4B) after the surgery proved that the vessels were perfusable and connected to the host vasculature, although the vessels were somewhat leaky (in this case without added bone marrow—derived MSC, bmMSC). The authors attributed the formation and (partial) stabilization of the blood vessels formed to the ability of the transplanted endothelial cells to elicit a beneficial host response that involved macrophages, fibroblasts, SMC, and host EC, and that eventually led to the remodeling of the implant and the formation of a pefusable vasculature, anastomosed to the host vascular system. In a later study using the same animal model, the authors embedded bmMSC inside the RAEC-coated collagen modules and succeeded in creating a less leaky, more mature vasculature, with the hematoma almost fully resolved by 21 days [4] (Figure 7.4C).

They reported significant remodeling of the prefabricated modular constructs in this case as well. At later time points (days 14 and 21), a high percentage of the implanted bmMSC were associated with the blood vessels formed in the implant area, forming the smooth muscle layer of these blood vessels. This study also showed that the bmMSC had a significant effect on the host macrophage response, with a decrease in the total number of macrophages (CD68+) apparent at day 14 and at day 21, and, most importantly, with an increase in the number of "pro-angiogenic" macrophages (CD163+) inside the implant area for modules embedded with bmMSC and coated with EC compared with modules without bmMSC. Thus, the addition of the bmMSC seemed to have a beneficial effect in terms of modulating the host response to the implanted modular tissue and creating a less leaky vasculature.

In an SCID/Bg mouse subcutaneous model with HUVEC-coated modules, modules were fabricated either using collagen alone or coating the collagen modules with fibronectin before seeding with HUVEC [17]. Fibronectin-coated modules decreased apoptosis of implanted HUVEC by nearly 40 percent (TUNEL staining) at day 3 after the surgery. This resulted in a nearly twofold increase in donor-derived (UEA-1+) blood vessel formation at day 7 and day 14 after surgery, although these differences did not persist at later time points (day 21).

HUVEC readily undergo apoptosis upon implantation, even in SCID/Bg mice [44], and this needs to be minimized (e.g., by using MSC or by genetic manipulation [44,45]) to drive vascular remodeling.

Adipose-derived MSC (adMSC) were embedded inside collagen modules, coated with human microvascular endothelial cells (HMEC), and implanted subcutaneously in an SCID/Bg mouse model [46]. In the absence of adMSC, there was limited HMEC survival past 14 days, whereas with the embedded adMSC, HMEC-derived primitive vessels formed as early as day 3, matured over time, connected to the host vasculature, and were still present in the implant area 90 days after surgery.

In turn, this early vascularization of the tissue was beneficial for fat development over time, with the implanted adMSC surviving (in the vascularized environment) and presumably differentiating over time into mature adipocytes. Approximately

60 percent of the implant showed fat accumulation (Oil Red O+) by 90 days (the duration of the study) when HMEC were included (Figure 7.4D). With modules embedded with adMSC but not coated with HMEC, there was limited fat accumulation over time, presumably due to the lack of early tissue vascularization and, consequently, early adMSC death. This study shows the advantages of this modular approach; while the adMSC initially contributed to HMEC survival and blood vessel formation (presumably due to adMSC secreted pro-angiogenic and pro-survival factors), this early tissue vascularization proved to be beneficial at later time points for adMSC survival and differentiation into what appears to be fat.

The creation of functional tissues using the modular tissue engineering approach has also been explored in vivo. An in vivo study investigated islet transplantation in the context of modular tissue engineering, using both syngeneic and allogeneic immunosuppressed diabetic rat omental pouch models [6]. In this case, the coating of the modules with EC resulted in increased blood vessel formation compared to transplantation of free islets or nonendothelialized islet modules. However, this increase in blood vessel density did not result in improved islet function (as determined through blood glucose and insulin measurements), presumably due to delayed blood vessel maturation and lack of initial functionality.

In an in vivo cardiac tissue engineering study, collagen modules (with or without Matrigel™) containing a rat neonatal cardiomyocyte-enriched cell population and coated with rat cardiac EC (both isolated from GFP transgenic animals) were implanted in syngeneic Lewis rats [47]. The modules were injected in the peri-infarct region of the heart 7 days after inducing myocardial infarction, and explanted and analyzed by histology 3 weeks after implantation. Donor-derived cardiomyocyte bundles (GFP+/myosin heavy chain [MHC]+) were observed at the implant site throughout the collagen modules for both EC-coated and non-EC-coated modules. EC coating of the collagen modules (without Matrigel™) was beneficial in terms of increasing total blood vessel density (donor and host-derived, CD31+) in the implant area compared with the cardiomyocyte-only implants (without EC coating). Some of these blood vessels contained erythrocytes, suggestive of connection to the host vasculature. Moreover, most of the blood vessels formed following implantation of the EC-coated modules contained donor-derived EC (GFP+). However, this increase in vascularization for the EC-coated modules did not lead to improved cardiomyocyte survival compared with the cardiomyocyte-only implants, at least not 3 weeks after implantation (time point included in this study).

In contrast to the collagen modules, modules made of a mixture of collagen and Matrigel™ performed less well in this animal model. The Matrigel™ containing implants elicited a much stronger macrophage (CD68+) and T-cell (T-cell receptor+) infiltration in the implant area, which presumably resulted in the loss of donor cardiomyocytes, emphasizing again the importance of modulating host response. No GFP+/MHC+ cardiomyocytes were detectable 3 weeks after implantation in the case of cardiomyocyte, EC-coated modules with Matrigel™. Some GFP + /MHC + cardiomyocytes were still found in cardiomyocyte-only implants, but mostly on the perimeter of the modules.

7.2.2 Cell sheet technology

Cell sheet stacking, as introduced by the Okano group, is another biofabrication technology that was used to build prevascularized modular tissue constructs for transplantation, with each cell sheet acting as a modular building block [48−50]. As has been demonstrated by others [51], cells may be cultured on thermoresponsive poly(N-isopropylacrylamide) (PNIPAAm)−based hydrogels at 37°C; then the temperature is decreased below the lower critical solution temperature, and the cells can be detached as a consequence of hydrogel swelling without enzymatic digestion. Using a hydrogel-coated plunger to harvest the cell sheet at the lower temperature, the first cell sheet is placed on top of a second cell sheet, the two cell sheets are allowed to attach, and the plunger is then lifted along with the double-layered cell sheet construct. The procedure is repeated to overlay multiple cell sheets and create thicker, 3D constructs (Figure 7.4E).

In one study aiming to create a prevascularized construct using the cell sheet technology, HUVEC sheets were sandwiched between myoblast cell sheets to obtain a five-layered construct [48]. The HUVEC formed tubular structures inside the construct after 4 days of culture in vitro. When these prevascularized multi-layered constructs were implanted subcutaneously in nude rats, some HUVEC-derived blood vessels (human CD31+ and UEA-1+) were visible 7 days after implantation. Some transplanted myoblasts (desmin+) were also present in the implant area at day 7 after implantation.

In a different study using a similar approach, EC were co-cultured with neonatal rat cardiomyocytes to create a prevascularized multilayered cardiac tissue [49]. In this study, each individual cell sheet contained both EC and cardiomyocytes, and three cell sheets were stacked together to build a larger construct. These three-layered constructs were grafted onto the infarcted myocardium of athymic rats 2 weeks after inducing myocardial infarction. Implants containing co-cultured EC and cardiomyocytes showed increased blood vessel formation (isolectin B4 staining), compared with implants with cardiomyocytes alone (without EC co-culture) at 4 weeks after implantation. The co-culture implants also reduced fibrosis compared with the cardiomyocyte-only implants. Cardiac function, as assessed by fractional shortening, was improved compared to sham, although only when higher EC densities ($\geq 4 \times 10^5$ EC/cell sheet) were implanted, but not for lower EC densities or cardiomyocyte-only implants; the end diastolic anterior wall thickness was increased compared to sham transplantation at the 4-week time point for all implants (with or without EC). The cell sheet technology was used in several other studies to improve cardiac function upon transplantation, using various other cell sources (skeletal myoblasts, mesenchymal stem cells), as reviewed elsewhere [52].

7.2.3 In vitro modular tissue engineering

Cardiac and liver tissues with an intrinsically vascularized structure are examples of tissue constructs built in vitro using a modular approach to include the presence of a vascular component as a biofabrication design requirement. In a cardiac modular tissue engineering study, a neonatal cardiomyocyte-enriched cell population was embedded in submillimeter-sized modules made of collagen and Matrigel™, and RAEC were seeded on the surface of the modules [5]. These modular units were then assembled into a macroporous sheet-like structures by gelling alginate on top of a single layer of modules to partially immobilize the modules together. The premise was that multiple sheets could ultimately be stacked together to create a 3D multisheet endothe-lialized cardiac tissue. Cardiac troponin I+ and connexin-43 + muscle bundles were observed in these modular structures, and the single module sheets were able to contract upon external field stimulation, although the presence of the EC coating on the modules interfered with the responsiveness of the constructs.

A modular approach has also been used to form vascularized liver tissue in vitro [53]. Primary rat hepatocytes were first cultured on a rotary shaker to form spheroids, then coated with collagen, and finally seeded with HUVEC. The endothelialized hepatocyte spheroids were then packed together inside hollow fibers (otherwise used for plasma separation). The hepatocytes retained liver-specific function (measured through albumin secretion), and endothelial cells were present inside the construct (vWF +) for the duration of the study (9 days in vitro).

Another in vitro study investigated the interaction of SMC and EC in the context of modular tissue engineering [3]. Depending on the presence or absence of serum in the culture medium, the authors found that the SMC phenotype changed and directly affected the phenotype of the co-cultured EC. SMC preconditioning in serum-free medium prior to embedding inside the modules improved EC-adherent junction formation on the surface of the modules (VE-cadherin expression, used as an indicator for EC quiescence) compared with SMC cultured in medium containing fetal bovine serum; however, it also increased HUVEC proliferation, thus emphasizing once more the importance of modulating the phenotype of each of the co-cultured cell types for a successful outcome. The goal had been to use SMC-EC co-culture as a means of maintaining an intact, quiescent, nonthrombogenic layer of EC on the surface of the modules. However, this goal has not been achieved using SMC as described above.

7.3 Building tissue constructs by assembling building blocks

Several groups used the modular approach as a bottom-up approach to build larger tissues from smaller building blocks in vitro, but without EC coating of the modules for vascularization. The final assembly of the modular units into a larger structure was done either in a random or a controlled manner if the objective was to control the architecture of the final construct.

7.3.1 Self-assembly of modules

Microtissues are submillimeter-sized cell aggregates (generally spheroids) that form by cellular self-assembly when cells are cultured under conditions that prevent their attachment to the culture dish surface and favor aggregation of adhesion-dependent cells. Microtissues are fabricated using different methods, such as culture on nonadherent surfaces, culture in hanging drops, culture in spinner flasks, and centrifugation-based methods [54,55]. They can serve as building blocks to form larger structures and can be assembled into defined shapes (Figure 7.5A).

The Morgan group investigated the in vitro parameters controlling cell aggregation within individual microtissues, as well as the self-assembly of multiple microtissues [56−62]. They showed that the duration of the preculture step before mixing the modular blocks together (hence the maturity of the microtissue building blocks) plays a particularly important role in the outcome of the self-assembly process [60]. When normal human fibroblast (NHF) spheroids were precultured for 1, 4, or 7 days and then assembled inside the recesses of through-shaped agarose gels, the spheroid modular blocks fused and assembled into a rod shape within 24 hours, regardless of the duration of the preculture step. However, the remodeling and self-assembly process was considerably slower for modules precultured for longer periods of time. On the other hand, the initial size of the module building blocks surprisingly had no effect on the kinetics of the assembly process or the length of the final rod structure, with similar results for both small and large modular building blocks. When mixing modules containing two different types of cells (NHF and H35 rat hepatoma cells), the position of each cell type within the final structure could be modulated by varying the preculture time of the individual building blocks. The structures obtained ranged from a single NHF core coated with H35 cells or several NHF spherical cores, but completely coated with H35 cells and fused together, or inside-out structures, and so on [60].

Furthermore, the assembled modules could be reassembled or remodeled by subsequent culture steps. When HUVEC spheroids precultured for 7 days were mixed with NHF cells, the HUVEC spheroids were able to reassemble and form microtissues with a completely distinct structure, with the NHF cells on the inside and coated with HUVEC on the outside [61]. The kinetics of cellular

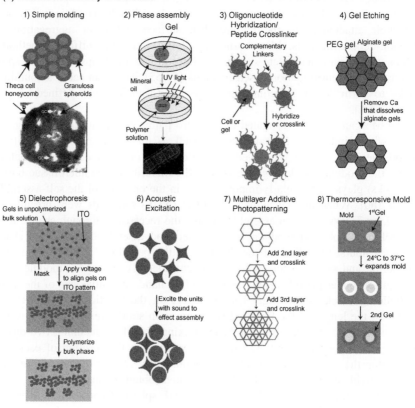

(a) Self-assembly of modules

Cells, spheroids, or
cells on microcarriers

Mold of shape (rod, toroid,
honeycomb, etc.)

Culture constructs
until units fuse

(b) Directed assembly of modules

1) Simple molding

Theca cell
honeycomb

Granulosa
spheroids

2) Phase assembly

Gel

Mineral
oil

UV light

Polymer
solution

3) Oligonucleotide
Hybridization/
Peptide Crosslinker

Complementary
Linkers

Cell or
gel

Hybridize
or crosslink

4) Gel Etching

PEG gel Alginate gel

Remove Ca
that dissolves
alginate gels

5) Dielectrophoresis

Gels in unpolymerized
bulk solution

ITO

Mask

Apply voltage
to align gels on
ITO pattern

Polymerize
bulk phase

6) Acoustic
Excitation

Excite the units
with sound to
effect assembly

7) Multilayer Additive
Photopatterning

Add 2nd layer
and crosslink

Add 3rd layer
and crosslink

8) Thermoresponsive Mold

Mold   1stGel

24°C to 37°C
expands mold

2nd Gel

FIGURE 7.5

(a) Self-assembly of modules. The self-assembly of modular units (cells, spheroids,
microcarriers, etc.) is driven by the aggregation of the individual units into a higher-order
shape (rod, toroid, honeycomb, etc.) within a mold. (b) Different methods of directed

(Continued)

◀ assembly of modular units. (1) Simple molding. A honeycomb pattern was formed from theca cells (an ovary stromal cell), and spheroids of granulosa cells were placed inside the voids in the honeycomb pattern. This formed an ovary-like structure. (2) Phase assembly uses two or more phases to thermodynamically drive the aggregation of the modular units together. After the units have aggregated, they are crosslinked again to hold the structure together. (3) Oligonucleotide hybridization and peptide crosslinkers. Individual units are formed and the surface is decorated with complementary linkers. These linkers can be oligonucleotides or peptides with free thiol groups. The units are aggregated, and conditions are changed to react the linkers together. (4) Gel etching uses units made from alginate and another polymer, such as PEG, to build the 3D architecture of the construct. After crosslinking the units together, the construct is washed with buffers to remove calcium from the gels. This causes the alginate gels to depolymerize, allowing channels and voids to be formed in the 3D construct. (5) Dielectrophoretic patterning uses a pattern laid out using a dielectric mask on indium tin oxide (ITO) to pattern gel microcarriers. The microcarriers are loaded into the device in an unpolymerized bulk solution of a polymer, such as PEG. A voltage is applied to the system that causes the microcarriers to migrate to the ITO surface, forming the pattern. The bulk phase polymer is then polymerized to lock the pattern of microcarriers in place. (6) Acoustic excitation uses sound waves to move different-shaped units to form complex repeating patterns. (7) Multilayer additive photopatterning uses a UV crosslinkable polymer as a mask to form a shape. The mask is then moved, and a new layer is added on top of the old layer. This process can be repeated several times. (8) Thermoresponsive mold. A mold is made from a substance that changes shape due to a temperature change such as PNIPAAm. A gel is formed in this mold at low temperature. The temperature is then increased to expand the mold. This causes a space to form around the first gel, which allows a second gel to be formed around the first.

Image reproduced from J Assist Reprod Genet [63], copyright 2010, with permission from Springer (A), pending permission (b) Image reproduced from J Assist Reprod Genet [63], copyright 2010, with permission from Springer (B1) , copyright 2011, with permission from John Wiley & Sons (B2) pending permission (c), image reproduced from Biotechnol Bioeng [28] pending permission (d)].

self-assembly also varied from one cell type to the other [57,62]. While H35 cells formed and maintained relatively stable rod, tori, or honeycomb structures inside the recesses of micromolded agarose gels, NHF cells were found to quickly reassemble the initial rod structures to a final spheroid structure, and the NHF cells also formed fewer stable tori and honeycomb structures as well [62].

In another study, multiple H35 toroid structures were randomly assembled together, with some of the toroids fused along their outer rim and giving rise to multilumen structures, while others were fused along their top and bottom surfaces, elongating the tubes. The authors suggested that the assembly of toroid modules might be useful in the context of tissue engineering for creating 3D porous structures with high cell density. Nevertheless, remodeling of the individual toroid structures over time and the eventual loss of the lumen due to remodeling still remain a concern. It was observed that the inner diameter of individual toroids significantly decreased over time (up to 85 percent decrease within 10 days, although it did not completely close, presumably due to the slower rate of self-assembly of H35 cells) [59]. Overall, these studies emphasize the importance of understanding self-assembly and tissue remodeling processes and perhaps finding ways to control or at least modulate these processes toward a successful outcome.

Exploiting another approach, self-assembly of fibroblast-coated gelatin microcarriers was used to build a dermal tissue equivalent [20]. Primary bovine fibroblasts were seeded onto gelatin microcarriers and cultured in a spinner flask, and these microtissues were then transferred to an assembly chamber. Histology and qRT-PCR analysis showed the presence of a newly synthesized layer of type I collagen (one of the main components of dermal tissue) inside and around these microtissues; the diameter of the beads also visibly increased over time, due to the synthesis and deposition of new ECM by the fibroblasts. After 1 week of culture and maturation in the assembly chamber, compact and homogeneous disc-shaped dermal tissue equivalents of about 1 cm diameter and 1 mm thick were obtained. The study showed that with this modular tissue engineering approach, fibroblasts were able to maintain their natural functions and quickly generated a dermal substitute that is rich in collagen type I and easy to handle without breaking.

In another study using gelatin microcarriers, human amniotic MSC were seeded onto gelatin microcarriers and induced to differentiate toward the osteogenic lineage within 28 days in spinner flask culture, with the goal of creating a bone substitute. The microtissues had increased mineral deposition over time (Alizarin red S staining) and also increased alkaline phosphatase (ALP) activity and increased mRNA expression of collagen type I and osteocalcin, indicative of osteogenic differentiation. These bone microtissues were then transferred to a perfusion chamber and cultured for 7 days under pulsatile flow with osteogenic culture medium, resulting in a larger bone tissue construct (approximately 2 cm diameter $\times$ 1 cm long) [21].

7.3.2 Directed assembly of modules

Much effort has also been devoted toward devising methods to control the assembly of the modules (Figure 7.5B). The premise is that by directing the assembly of the modular building blocks (instead of randomly packing multiple modules together), it will be possible to recreate native tissue architecture, as well as the native cell—cell interactions between the multiple cell types present in a tissue, with the overall goal of building better mimics of native tissues. Nevertheless, remodeling of these preformed in vitro structures upon implantation in vivo is a question still requiring further investigation, as these preformed modular assemblies have been generally characterized only in vitro so far.

7.3.2.1 Manual assembly of functional tissue components

The functional benefit of modular assembly was eloquently demonstrated in an in vitro study aiming to fabricate a human ovary mimic [63]. The "human artificial ovary" was assembled from constituent modular blocks containing the three functional cell types that are naturally found in an ovarian follicle—namely, the theca cells, the granulosa cells, and the oocytes. Theca cells were first placed in an agarose mold containing honeycomb-shaped wells. Once formed (within 1 day), the honeycomb-shaped theca cell microtissues were then removed from the agarose mold and placed in a petri dish. Cumulus granulosa-oocyte complexes (COC) were then transferred inside the openings of the honeycombs formed by the theca cells. Within a few hours, the theca cells contracted and immobilized the COC inside the construct. One of three oocytes cultured using this method and extracted from the construct at 45 hours demonstrated polar body extrusion, which indicated maturation of the oocyte. Within a few days, the microtissue remodeled to a spheroid shape containing all three cell types. The authors concluded that the suggested assembly and co-culture of the three types of cells were beneficial for oocyte maturation, demonstrating the potential of a 3D modular assembly and the importance of cell interactions between the three functional cell types in order to create a successful in vitro maturation culture system or an in vitro toxicology model for the ovarian follicle.

7.3.2.2 Thermodynamically driven assembly in multiphase systems

The Khademhosseini group developed several methods of directed module assembly based on the thermodynamic tendency of multiphase systems to minimize the contact surface area between phases [26−29,64,65]. These methods rely on the tendency of hydrogels (with hydrophilic properties) to pack together in the presence of a hydrophobic phase in order to minimize the surface of contact with the hydrophobic phase. By varying the different fabrication parameters, such as the initial size and shape of the hydrogels, it is possible to build different modular structures, with lock-and-key shapes being particularly well suited for controlling the assembly of the building blocks. In one study, PEG methacrylate hydrogels were fabricated by photolithography and then placed in a petri dish containing mineral oil. Using a pipette tip, the modules were manually agitated inside the petri dish, which led to module aggregation. To preserve the structure of the modular assembly, the modules were reexposed to UV light prior to removal from the oil phase (secondary UV crosslinking step). Most of the NIH/3T3 fibroblasts encapsulated inside the hydrogels survived the assembly process (Live/Dead® assay immediately after each biofabrication step), although both the agitation and the UV photocrosslinking steps lowered the cell viability somewhat [26].

Using the same method, double-layered, concentric tubular constructs that mimic the structure of a blood vessel were assembled [28]. EC were encapsulated in the internal ring, and SMC were encapsulated in the external ring of each PEG hydrogel module (two-step photolithography); then several modules were assembled together in oil to form a double-layered tubular structure. This proof-of-concept study showed that the layered tubular structure of a blood vessel can be recreated using this assembly approach. However, since PEG has low cell attachment properties, and it is not degradable, it is expected that the long-term survival of the cells within this matrix will be limited and the use of a more bioactive hydrogel material is likely necessary for future long-term studies.

Using a similar principle, hydrogels floating on the surface of high-density hydrophobic liquids such as perfluorodecalin (PFDC) or carbon tetrachloride (CCl_4) were driven by the surface tension on the liquid—air interface to move toward each other and aggregate [29]. This method was used to assemble together modules containing two different cell types (HepG2 and NIH/3T3), such that each HepG2 module was surrounded by six NIH/3T3 modules. Gelatin methacrylate was used instead of PEG to fabricate these hydrogels to allow cell attachment and migration. As an alternative to module assembly in hydrophobic liquids, hydrogel modules were also placed on glass surfaces patterned with hydrophobic and hydrophilic regions. The glass slides were patterned with hydrophobic regions using PDMS stamps coated with octadecyltrichlorosilane (OTS). The modules had a tendency to assemble onto the hydrophilic regions of the glass slides, driven by surface tension [64]. The assembled modular sheets were then stabilized through a secondary UV photocrosslinking step.

7.3.2.3 Assembly through chemical approaches

Gartner and Bertozzi suggested a method to control the connectivity between cells; their method involved functionalization of the cells with short DNA sequences, followed by formation of controlled cell—cell contacts through hybridization of complementary DNA sequences and purification of the desired cellular structures from the byproducts or unreacted components by fluorescence-activated cell sorting [66]. This method of assembly controls the architecture of the microtissues at the cellular level, with each individual cell becoming a building block. The authors used this assembly method as a proof-of-concept to build a paracrine signaling network in a 3D environment in vitro. CHO cells engineered to express interleukin-3 (IL-3) and a hematopoietic progenitor cell line (FL5.12) that undergoes apoptosis in the absence of IL-3 were functionalized with complementary oligoneuclotides and assembled together. The assembled microtissues were then embedded in an agarose gel. The study showed that FL5.12 cells that had formed oligonucleotide-mediated cell—cell contacts with the CHO cells expressing IL-3 were indeed able to survive and proliferate, but not FL5.12 that were not part of a CHO/FL5.12 module, thus confirming the formation of a signaling network between CHO and FL5.12 cells connected through complementary oligonucleotides.

Using a related directed assembly approach based on hybridization of complementary DNA sequences, the Bhatia group functionalized PEG diacrylate photocrosslinked hydrogels with DNA sequences [67]. The PEG microgels were first decorated with streptavidin moieties, which were then linked to biotinylated DNA sequences. The assembly of the microgels was directed through hybridization of the DNA sequences present on the surface of the microgels to the complementary DNA sequences spotted in a controlled fashion onto a DNA array template. It was shown that this microgel-directed assembly method is applicable to multiple cell types (J2-3T3 rat fibroblasts, TK6 human lymphoblasts, A549 human lung adenocarcinoma cells were encapsulated inside the hydrogels). However, only immediate survival (3 hours after encapsulation) was demonstrated, and the long-term survival and functionality of the microtissues assembled using this method remain to be investigated in future studies.

Another method of assembly of PEG hydrogels relies on using a peptide cross-linker with free thiol groups to bridge the PEG diacrylate hydrogels together (Michael-type addition reaction) [68]. PEG hydrogels of various shapes (squares, circles, stars) were fabricated and assembled using this method, with star-shaped hydrogel assemblies having the highest porosity and pore interconnectivity. NIH/3T3 fibroblasts embedded inside star-shaped hydrogels and assembled using this method were able to survive for at least 3 days under perfusion culture in vitro. Long-term survival studies, as well as functional assays, potentially with thera-peutic cells of interest, remain to be investigated using this biofabrication method.

7.3.2.4 *Porous assembly using sacrificial building blocks*

Porous 3D modular constructs were also fabricated by stacking several layers of PEG-based modules together and using alginate microgels as sacrificial building blocks to create pores within the construct. Modules were first fabricated by photocrosslinking a PEG diacrylate solution inside a PDMS mold. The hydrogels were then retrieved and assembled in a monolayer structure, further stabilized by a secondary UV crosslinking step. Multiple layers were then stacked together to create a larger 3D construct and stabilized through further photocrosslinking [69]. While NIH 3T3 cells remained about 80 percent viable in all four layers of the multilayered construct, HepG2 cells were only about 70 percent viable in the top and bottom layer 24 hours after fabrication, potentially due to the toxicity of photocrosslinking and the lack of bioactive components in PEG. Most impor-tantly, the majority of HepG2 in the middle layer of the three-layered construct did not survive (less than 15 to 20 percent survival after 24 hours in culture). In some cases, alginate microgels were mixed with the PEG microgels and dissolved after assembly to create pores within the final structure. Live/Dead® fluorescence images showed that the initial survival (24 hours) of HepG2 in the middle layer was improved when porous structures were created using sacrificial alginate microgels, presumably due to better oxygen and nutrient accessibility through the pores. Future studies are needed to evaluate the survival of the cells at later time points, and functional assays are required to test the suitability of the construct for tissue engineering.

7.3.2.5 *Assembly through microfluidic approaches*

Another method of directed assembly of modular building blocks uses dielectro-phoretic (DEP) patterning [70]. Hydrogel microtissues with encapsulated cells are mixed with a bulk-phase polymer solution of low conductivity and low viscosity (amenable to DEP patterning) and placed inside a DEP patterning chamber. Micropatterning of a dielectric layer on the lower glass slide of the DEP chamber allows for directed positioning of the microgels in the areas of the DEP chamber with higher electric field strength. Once the microgels are assembled in the desired regions of the chamber, the bulk-phase polymer is gelled either through exposure to UV light or a change of temperature to preserve the microgel assembly. The whole construct can then be transferred to a cell culture dish. This directed assembly biofabrication method has been used with bipotential mouse embryonic liver cells encapsulated in alginate microgels and then mixed with an agarose solution. After positioning the microgels inside the agarose solution, the agarose was gelled by brief immersion in ice water to stabilize the microgel assembly. Most BMEL cells survived the biofabrication process (Live/Dead® assay performed 1 hour after polymerization), although long-term viability and functional assays remain to be performed in future studies to confirm the biocompatibility of this method.

Acoustic excitation was also used as a method to assemble microgel modules of different sizes and shapes (cubes, tetris shapes, saw-like shapes, lock-and-key shapes) [30]. PEG-based microgels were fabricated using common photolithography methods and then assembled by exposure to acoustic excitation inside a petri dish assembly chamber. The assembly process was complete within seconds. Several layers were also assembled using this method by first photocrosslinking the first layer of assembled modules, followed by step-by-step assembly and photocrosslinking of each subsequent layer. Cells encapsulated in the microgels were viable immediately after the assembly process (Live/Dead® assay performed on NIH/3T3 fibroblasts immediately after acoustic assembly). However, the long-term and functional performance of these constructs remains to be evaluated.

A different technique—multilayer additive photopatterning—has also been used to build modular tissues with controlled architecture [31]. The individual layers (which served as modular building blocks in this case) were made through placement under a photomask (of the desired shape) and UV photocrosslinking of a PEG-based polymer solution functionalized with RGD cell attachment sequences. The polymer solution that was not crosslinked during the first step of photopatterning was then washed away, and a second layer of PEG-based polymer was added using a thicker spacer and photocrosslinked using a new photomask. The procedure was repeated to obtain the third layer. Primary hepatocytes were encapsulated in each of the layers of the construct to obtain a 3D multilayered hepatic tissue. The design of the photomasks ensured that none of the regions of the cell polymer mixture was exposed to UV light more than once, thus enhancing cell viability. By assembling the three layers, a hexagonal branching structure was obtained. An MTT assay performed 24 hours after cell encapsulation within a single-layer construct (hexagon shape with photopatterned features of 500 μm) showed that hepatocyte short-term survival was improved in photopatterned hydrogels compared to unpatterned bulk hydrogels that exhibited a necrotic core. The necrotic core was presumably due to poor oxygen and nutrient transport at the center of the unpatterned hydrogel discs. Three layered hexagonal branched constructs cultured for 12 days in vitro under perfusion produced more albumin and urea compared to unpatterned hydrogels, indicators of improved functionality of the engineered liver tissue when fabricated using the multilayer modular approach.

Two-layered, sequentially patterned modules were also formed using thermo-responsive PNIPAAm micromolds [71]. Using the swelling or shrinking properties of the PNIPAAm polymer below or above its lower critical solution temperature (~32°C), the shape of the micromold was dynamically adjusted by changing the temperature, and two separate agarose layers (in some cases with encapsulated cells) were sequentially gelled inside the mold. The second agarose gel and cell population surrounded the original hydrogel layer and formed a two-layered construct with cylindrical, square or stripe shapes, depending on the shape of the micromold.

Microfluidics was also combined with modular tissue engineering with the purpose of fabricating large numbers of modules in a continuous and rapid manner. Using stop-flow lithography, microgels were formed by flowing a PEG diacrylate polymer solution through a microchannel and photocrosslinking the polymer solution by passing pulses of UV light through a photomask [72]. Large numbers of microgels of different shapes were fabricated using different photomasks. However, the viability of NIH 3T3 fibroblast cells encapsulated in the hydrogels fabricated using SFL was only 68 percent at 1 hour after fabrication, and further improvements to this method are likely necessary to make it more widely applicable.

7.4 Modular tissue engineering combined with microfluidics

Beyond what has been described in the previous section, microfluidic systems have been used by a few groups in the context of modular tissue engineering as a means of creating controlled 3D co-culture systems and in vitro tissue models. These systems combine the high-throughput and controlled microenvironment advantages of microfluidic devices with the advantages of modular tissue engineering, such as 3D cell culture and potential to culture multiple cell types together by mixing modules encapsulating different cell populations.

In a proof-of-concept study, collagen modules with embedded NIH/3T3 fibroblasts or HepG2 cells were fabricated inside a PDMS mold, loaded inside a microfluidics chamber, and cultured under perfusion for 24 hours [73]. The pores formed between the modules allowed for culture medium to circulate in the spaces between the modules and reach the encapsulated cells. The cells maintained high viability over 24 hours in perfusion culture (~99 percent). Narrowing the channels within the microfluidic chip allowed for ordered packing of the modules within the device. Sequentially loading three different batches of modules, each batch containing a differently labeled population of NIH/3T3 fibroblasts, resulted in an organized string of distinct modules. This technique is thus presumably amenable to the formation and culture of ordered modular tissue structures within the microfluidic chamber. Nevertheless, a material that is more amenable to perfusion culture (better mechanical properties) than collagen would presumably need to be used, as the authors noticed that the weak collagen modules deformed under flow, which limited the flow rate as well as the maximum length of the construct that could be perfused to only a few millimeters.

In a different study combining modular tissue engineering and microfluidics, a microfluidic chamber loaded with EC-coated collagen modules was used to characterize the remodeling of the modular construct and the changes in EC phenotype under flow conditions [74]. The study found that EC proliferation was significantly reduced for flow compared to static conditions after 24 hours (BrdU uptake assay), but similarly low proliferation levels were observed for both cases after 48 hours. KLF2 (a transcription factor upregulated with laminar shear stress and that upregulates antithrombotic factors) was upregulated at 24 hours under flow. However, expression of VE-cadherin (used as an indicator of EC quiescence) was downregulated and discontinuous under flow conditions, presumably due to EC reorientation and remodeling of the construct under flow. No statistically significant differences in VCAM-1 or ICAM-1 expression (indicators of EC activation) were observed between the static and flow conditions over 24 hours. Overall, the study suggested that perfusion of the modular constructs did not significantly increase EC activation; however, remodeling of the constructs under flow occurred, with discontinuous VE-cadherin expression, and with the void spaces (flow channels) formed in between the modules decreasing in size over time due to compaction, even at low flow rates.

When bmMSC were embedded inside the collagen modules prior to EC seeding, and using the same microfluidic remodeling chamber experimental setting, it was noted that flow conditioning of the modular constructs led to increased expression of smooth muscle cell markers (SMA+, desmin+ staining) by the MSC and increased migration from the inside of the modules toward the module surface compared with the static control, as well as compared with the MSC only (no EC coating) control, over the 21-day course of the study [75]. All modules also remodeled and contracted significantly over time, with the least amount of contraction observed for the MSC-only modules under static conditions. A change in the extracellular matrix composition was also observed for MSC-containing modules under flow conditions, with or without the EC coating, with more proteoglycan deposition occurring in these cases (Alcian blue staining). Overall, this study demonstrated the potential usefulness of these flow-conditioned modular tissues as model systems to study the remodeling of tissue engineered constructs composed of multiple cell types over time in vitro and as a tool to better understand and predict the remodeling that would occur upon implantation in vivo.

Other studies combined microfluidics with a modular approach to create in vitro 3D liver tissue models for drug metabolism and toxicology assays [76,77]. Several precision-cut rat liver slices were embedded in Matrigel™, placed inside a PDMS microfluidic device, and cultured under perfusion using an optimized culture medium formulation [77]. Each of the precision-cut liver slices (modular building blocks) maintains the native liver architecture and cellular components, making the system a closer mimic to the native tissue. Embedding the liver slices in Matrigel™ and using an optimized culture medium composition maintained high viability of the liver slices for at least 72 hours in culture (~90 percent by lactate dehydrogenase [LDH] activity assay), and preserved the liver metabolic activity (determined through measurement of metabolite formation upon addition of substrate), although only partially. Encouragingly, phase II metabolism was similar between fresh slices and slices embedded in Matrigel™ and cultured inside the microfluidic device for 72 hours; however, the phase I metabolism decreased significantly, with only 5 percent of the initial value after 72 hours.

CONCLUSION

Biofabrication of tissue constructs using a modular approach offers tremendous opportunities in terms of creating tissue constructs that recapitulate the complexity of native tissues. This in turn translates into more realistic in vitro models for drug testing or other biological assays, compared to conventional 2D systems or bulk scaffolds seeded with cells. For transplantation purposes, it also brings us a step closer toward creating tissue or organ replacements that are better mimics of their native counterparts.

Using the modular approach, multiple cell types can be mixed together, either within the same modular unit or in separate modules, and the relative abundance of each cell type can be individually controlled by simply changing the number of modules containing that particular cell population. Furthermore, the modular approach is 3D in nature, and it is versatile in terms of the materials that can be used to fabricate the modules, thus offering the opportunity to recreate the cells' natural ECM and microenvironment. Moreover, the porous structures created upon assembly of several modules are amenable to perfusion, enabling researchers to conduct biological assays under physiologically relevant flow conditions. The modular approach is also scalable, paving the way to fabricating tissues and organs of clinically relevant size. Increasingly larger structures, with cells evenly distributed throughout the construct, can be obtained by simply increasing the number of modular building blocks mixed together and controlling the cell density within each individual building block.

Moving forward, the question of remodeling of these preformed modular structures, particularly in vivo, needs to be further investigated. Host response and remodeling play a critical role in determining the success or failure of the implanted tissue construct. Therefore, the in vivo remodeling of the engineered tissue needs to be integrated as a design parameter for the fabrication of these structures in vitro. Finding ways to modulate the remodeling of the tissue constructs after implantation is perhaps more important than achieving perfect control over the tissue microarchitecture prior to implantation. Distinct sets of design criteria may in fact be necessary depending on whether the envisioned application is fabrication of tissue or organ replacements for transplantation, in which case host response and in vivo remodeling are probably the key design parameters to consider or whether the envisioned application is fabrication of in vitro tissue models, in which case faithful imitation of the native tissue structure may be of highest importance.

References

[1] McGuigan AP, Sefton MV. Vascularized organoid engineered by modular assembly enables blood perfusion. Proc Natl Acad Sci USA 2006;103:11461—6.

[2] McGuigan AP, Leung B, Sefton MV. Fabrication of cell-containing gel modules to assemble modular tissue-engineered constructs [corrected]. Nat Protoc 2006;1: 2963—9.

[3] Leung BM, Sefton MV. A modular tissue engineering construct containing smooth muscle cells and endothelial cells. Ann Biomed Eng 2007;35:2039—49.

[4] Chamberlain MD, Gupta R, Sefton MV. Bone marrow-derived mesenchymal stromal cells enhance chimeric vessel development driven by endothelial cell-coated microtissues. Tissue Eng Part A 2012;18:285—94.

[5] Leung BM, Sefton MV. A modular approach to cardiac tissue engineering. Tissue Eng Part A 2010;16:3207—18.

[6] Gupta R, Sefton MV. Application of an endothelialized modular construct for islet transplantation in syngeneic and allogeneic immunosuppressed rat models. Tissue Eng Part A 2011;17:2005—15.

[7] Carmeliet P, Jain RK. Angiogenesis in cancer and other diseases. Nature 2000;407:249—57.

[8] Corstorphine L, Sefton MV. Effectiveness factor and diffusion limitations in collagen gel modules containing HepG2 cells. J Tissue Eng Regen Med 2011;5:119—29.

[9] Gauvin R, Khademhosseini A. Microscale technologies and modular approaches for tissue engineering: moving toward the fabrication of complex functional structures. ACS Nano 2011;5:4258—64.

[10] Nichol JW, Khademhosseini A. Modular tissue engineering: engineering biological tissues from the bottom up. Soft Matter 2009;5:1312—9.

[11] Zorlutuna P, Annabi N, Camci-Unal G, Nikkhah M, Cha JM, Nichol JW, et al. Microfabricated biomaterials for engineering 3D tissues. Adv Mater 2012;24:1782—804.

[12] Elbert DL. Bottom-up tissue engineering. Curr Opin Biotechnol 2011;22:674—80.

[13] Choudhury D, Mo X, Iliescu C, Tan LL, Tong WH, Yu H. Exploitation of physical and chemical constraints for three-dimensional microtissue construction in microfluidics. Biomicrofluidics 2011;5:22203.

[14] Rivron NC, Rouwkema J, Truckenmuller R, Karperien M, De Boer J, van Blitterswijk CA. Tissue assembly and organization: developmental mechanisms in microfabricated tissues. Biomaterials 2009;30:4851−8.

[15] Chamberlain MD, Butler MJ, Ciucurel EC, Fitzpatrick LE, Khan OF, Leung BM, et al. Fabrication of micro-tissues using modules of collagen gel containing cells. J Vis Exp 2010;46:e2177.

[16] Voice DN. High-throughput modular tissue engineering and applications to scale-up tissue constructs. M.A.Sc. thesis, University of Toronto; 2012.

[17] Cooper TP, Sefton MV. Fibronectin coating of collagen modules increases in vivo HUVEC survival and vessel formation in SCID mice. Acta Biomater 2011;7: 1072−83.

[18] McGuigan AP, Bruzewicz DA, Glavan A, Butte MJ, Whitesides GM. Cell encapsulation in sub-mm sized gel modules using replica molding. PLoS One 2008;3:e2258.

[19] Xu F, Moon SJ, Emre AE, Turali ES, Song YS, Hacking SA, et al. A droplet-based building block approach for bladder smooth muscle cell (SMC) proliferation. Biofabrication 2010;2:014105.

[20] Palmiero C, Imparato G, Urciuolo F, Netti P. Engineered dermal equivalent tissue in vitro by assembly of microtissue precursors. Acta Biomater 2010;6:2548−53.

[21] Chen M, Wang X, Ye Z, Zhang Y, Zhou Y, Tan WS. A modular approach to the engineering of a centimeter-sized bone tissue construct with human amniotic mesenchymal stem cells-laden microcarriers. Biomaterials 2011;32:7532−42.

[22] Nichol JW, Koshy ST, Bae H, Hwang CM, Yamanlar S, Khademhosseini A. Cell-laden microengineered gelatin methacrylate hydrogels. Biomaterials 2010;31:5536−44.

[23] Shin SR, Bae H, Cha JM, Mun JY, Chen YC, Tekin H, et al. Carbon nanotube reinforced hybrid microgels as scaffold materials for cell encapsulation. ACS Nano 2012;6: 362−72.

[24] Bae H, Ahari AF, Shin H, Nichol JW, Hutson CB, Masaeli M, et al. Cell-laden microengineered pullulan methacrylate hydrogels promote cell proliferation and 3D cluster formation. Soft Matter 2011;7:1903−11.

[25] Yeh J, Ling Y, Karp JM, Gantz J, Eng G, Chandawarkar A, et al. Micromolding of shape-controlled, harvestable cell-laden hydrogels. Biomaterials 2006;27:5391−8.

[26] Du Y, Lo E, Ali S, Khademhosseini A. Directed assembly of cell-laden microgels for fabrication of 3D tissue constructs. Proc Natl Acad Sci USA 2008;105: 9522−7.

[27] Du Y, Lo E, Vidula MK, Khabiry M, Khademhosseini A. Method of bottom-up directed assembly of cell-laden microgels. Cell Mol Bioeng 2008;1:157−62.

[28] Du Y, Ghodousi M, Qi H, Haas N, Xiao W, Khademhosseini A. Sequential assembly of cell-laden hydrogel constructs to engineer vascular-like microchannels. Biotechnol Bioeng 2011;108:1693−703.

[29] Zamanian B, Masaeli M, Nichol JW, Khabiry M, Hancock MJ, Bae H, et al. Interface-directed self-assembly of cell-laden microgels. Small 2010;6:937−44.

[30] Xu F, Finley TD, Turkaydin M, Sung Y, Gurkan UA, Yavuz AS, et al. The assembly of cell-encapsulating microscale hydrogels using acoustic waves. Biomaterials 2011;32:7847−55.

[31] Liu Tsang V, Chen AA, Cho LM, Jadin KD, Sah RL, DeLong S, et al. Fabrication of 3D hepatic tissues by additive photopatterning of cellular hydrogels. FASEB J 2007;21: 790−801.

[32] Sosnik A, Leung B, McGuigan AP, Sefton MV. Collagen/poloxamine hydrogels: cytocompatibility of embedded HepG2 cells and surface-attached endothelial cells. Tissue Eng 2005;11:1807—16.

[33] Sosnik A, Sefton MV. Semi-synthetic collagen/poloxamine matrices for tissue engineering. Biomaterials 2005;26:7425—35.

[34] Sosnik A, Sefton MV. Poloxamine hydrogels with a quaternary ammonium modification to improve cell attachment. J Biomed Mater Res A 2005;75:295—307.

[35] Sosnik A, Sefton MV. Methylation of poloxamine for enhanced cell adhesion. Biomacromolecules 2006;7:331—8.

[36] Ciucurel EC, Sefton MV. A poloxamine-polylysine acrylate scaffold for modular tissue engineering. J Biomater Sci Polym Ed 2010;22:2515—28.

[37] Butler MJ. Modular approach to adipose tissue engineering [PhD thesis], University of Toronto; 2010.

[38] Sosnik A, Leung BM, Sefton MV. Lactoyl-poloxamine/collagen matrix for cell-containing tissue engineering modules. J Biomed Mater Res A 2008;86:339—53.

[39] Hutson CB, Nichol JW, Aubin H, Bae H, Yamanlar S, Al-Haque S, et al. Synthesis and characterization of tunable poly(ethylene glycol): gelatin methacrylate composite hydrogels. Tissue Eng Part A 2011;17:1713—23.

[40] McGuigan AP, Sefton MV. The thrombogenicity of human umbilical vein endothelial cell seeded collagen modules. Biomaterials 2008;29:2453—63.

[41] She M, McGuigan AP, Sefton MV. Tissue factor and thrombomodulin expression on endothelial cell-seeded collagen modules for tissue engineering. J Biomed Mater Res A 2007;80:497—504.

[42] Gupta R, Van Rooijen N, Sefton MV. Fate of endothelialized modular constructs implanted in an omental pouch in nude rats. Tissue Eng Part A 2009;15:2875—87.

[43] Chamberlain MD, Gupta R, Sefton MV. Chimeric vessel tissue engineering driven by endothelialized modules in immunosuppressed Sprague-Dawley rats. Tissue Eng Part A 2011;17:151—60.

[44] Schechner JS, Nath AK, Zheng L, Kluger MS, Hughes CC, Sierra-Honigmann MR, et al. In vivo formation of complex microvessels lined by human endothelial cells in an immunodeficient mouse. Proc Natl Acad Sci USA 2000;97:9191—6.

[45] Enis DR, Shepherd BR, Wang Y, Qasim A, Shanahan CM, Weissberg PL, et al. Induction, differentiation, and remodeling of blood vessels after transplantation of Bcl-2-transduced endothelial cells. Proc Natl Acad Sci USA 2005;102:425—30.

[46] Butler MJ, Sefton MV. Cotransplantation of adipose-derived mesenchymal stromal cells and endothelial cells in a modular construct drives vascularization in SCID/bg mice. Tissue Eng Part A 2012;18:1628—41.

[47] Leung BM, Miyagi Y, Li RK, Sefton MV. Fate of modular cardiac tissue constructs in a syngeneic rat model. J Tissue Eng Regen Med [in press].

[48] Sasagawa T, Shimizu T, Sekiya S, Haraguchi Y, Yamato M, Sawa Y, et al. Design of prevascularized three-dimensional cell-dense tissues using a cell sheet stacking manipulation technology. Biomaterials 2010;31:1646—54.

[49] Sekine H, Shimizu T, Hobo K, Sekiya S, Yang J, Yamato M, et al. Endothelial cell coculture within tissue-engineered cardiomyocyte sheets enhances neovascularization and improves cardiac function of ischemic hearts. Circulation 2008;118: S145—52.

[50] Haraguchi Y, Shimizu T, Sasagawa T, Sekine H, SakaguchI K, Kikuchi T, et al. Fabrication of functional three-dimensional tissues by stacking cell sheets in vitro. Nat Protoc 2012;7:850−8.

[51] Rollason G, Davies JE, Sefton MV. Preliminary report on cell culture on a thermally reversible copolymer. Biomaterials 1993;14:153−5.

[52] Sekine H, Shimizu T, Okano T. Myocardial tissue engineering: toward a bioartificial pump. Cell Tissue Res 2012;347:775−82.

[53] Inamori M, Mizumoto H, Kajiwara T. An approach for formation of vascularized liver tissue by endothelial cell-covered hepatocyte spheroid integration. Tissue Eng Part A 2009;15:2029−37.

[54] Kelm JM, Fussenegger M. Microscale tissue engineering using gravity-enforced cell assembly. Trends Biotechnol 2004;22:195−202.

[55] Kelm JM, Fussenegger M. Scaffold-free cell delivery for use in regenerative medicine. Adv Drug Deliv Rev 2010;62:753−64.

[56] Youssef J, Nurse AK, Freund LB, Morgan JR. Quantification of the forces driving self-assembly of three-dimensional microtissues. Proc Natl Acad Sci USA 2011;108:6993−8.

[57] Tejavibulya N, Youssef J, Bao B, Ferruccio TM, Morgan JR. Directed self-assembly of large scaffold-free multi-cellular honeycomb structures. Biofabrication 2011; 3:034110.

[58] Bao B, Jiang J, Yanase T, Nishi Y, Morgan JR. Connexon-mediated cell adhesion drives microtissue self-assembly. FASEB J 2011;25:255−64.

[59] Livoti CM, Morgan JR. Self-assembly and tissue fusion of toroid-shaped minimal building units. Tissue Eng Part A 2010;16:2051−61.

[60] Rago AP, Dean DM, Morgan JR. Controlling cell position in complex heterotypic 3D microtissues by tissue fusion. Biotechnol Bioeng 2009;102:1231−41.

[61] Napolitano AP, Chai P, Dean DM, Morgan JR. Dynamics of the self-assembly of complex cellular aggregates on micromolded nonadhesive hydrogels. Tissue Eng 2007;13:2087−94.

[62] Dean DM, Napolitano AP, Youssef J, Morgan JR. Rods, tori, and honeycombs: the directed self-assembly of microtissues with prescribed microscale geometries. FASEB J 2007;21:4005−12.

[63] Krotz SP, Robins JC, Ferruccio TM, Moore, R, Steinhoff MM, Morgan JR, et al. In vitro maturation of oocytes via the pre-fabricated self-assembled artificial human ovary. J Assist Reprod Genet 27:743−750.

[64] Du Y, Ghodousi M, Lo E, Vidula MK, Emiroglu O, Khademhosseini A. Surface-directed assembly of cell-laden microgels. Biotechnol Bioeng 2010;105:655−62.

[65] Shi Z, Chen N, Du Y, Khademhosseini A, Alber M. Stochastic model of self-assembly of cell-laden hydrogels. Phys Rev E Stat Nonlin Soft Matter Phys 2009;80:061901.

[66] Gartner ZJ, Bertozzi CR. Programmed assembly of 3-dimensional microtissues with defined cellular connectivity. Proc Natl Acad Sci U S A 2009;106:4606−10.

[67] Li CY, Wood DK, Hsu CM, Bhatia SN. DNA-templated assembly of droplet-derived PEG microtissues. Lab Chip 2011;11:2967−75.

[68] Liu B, Liu Y, Lewis AK, Shen W. Modularly assembled porous cell-laden hydrogels. Biomaterials 2010;31:4918−25.

[69]　Yanagawa F, Kaji H, Jang YH, Bae H, Yanan D, Fukuda J, et al. Directed assembly of cell-laden microgels for building porous three-dimensional tissue constructs. J Biomed Mater Res A 2011;97A:93−102.

[70]　Albrecht DR, Underhill GH, Mendelson A, Bhatia SN. Multiphase electropatterning of cells and biomaterials. Lab Chip 2007;7:702−9.

[71]　Tekin H, Tsinman T, Sanchez JG, Jones BJ, Nichol JW, Camci-Unal G, et al. Responsive micromolds for sequential patterning of hydrogel microstructures. J Am Chem Soc 2011;133:12944−7.

[72]　Panda P, Ali S, Lo E, Chung BG, Hatton TA, Khademhosseini A, et al. Stop-flow lithography to generate cell-laden microgel particles. Lab Chip 2008;8:1056−61.

[73]　Bruzewicz DA, McGuigan AP, Whitesides GM. Fabrication of a modular tissue construct in a microfluidic chip. Lab Chip 2008;8:663−71.

[74]　Khan OF, Sefton MV. Perfusion and characterization of an endothelial cell-seeded modular tissue engineered construct formed in a microfluidic remodeling chamber. Biomaterials 2010;31:8254−61.

[75]　Khan OF, Chamberlain MD, Sefton MV. Toward an in vitro vasculature: differentiation of mesenchymal stromal cells within an endothelial cell-seeded modular construct in a microfluidic flow chamber. Tissue Eng Part A 2012;18:744−56.

[76]　Van Midwoud PM, Groothuis GM, Merema MT, Verpoorte E. Microfluidic biochip for the perifusion of precision-cut rat liver slices for metabolism and toxicology studies. Biotechnol Bioeng 2010;105:184−94.

[77]　Van Midwoud PM, Merema MT, Verweij N, Groothuis GM, Verpoorte E. Hydrogel embedding of precision-cut liver slices in a microfluidic device improves drug metabolic activity. Biotechnol Bioeng 2011;108:1404−12.

Formation of Multicellular Microtissues and Applications in Biofabrication

8

Andrew M. Blakely, Jacquelyn Y. Schell, Adam P. Rago, Peter R. Chai, Anthony P. Napolitano, Jeffrey R. Morgan

Department of Molecular Pharmacology, Physiology, and Biotechnology, Center for Biomedical Engineering, Brown University, Providence, Rhode Island, USA

CONTENTS

INTRODUCTION

Three-dimensional (3D) cell culture that more closely emulates in vivo tissue architecture and function is growing in importance in the fields of regenerative medicine, tissue engineering, and biofabrication. Tissue engineering encompasses all processes that involve culturing and directing cells to create an in vitro tissue that replicates the in vivo functions of tissues. The ultimate goal of tissue engineering is to grow, build, or biofabricate a tissue that can be transplanted to replace impaired or lost tissue function in the host. Applications for these 3D cell culture systems are emerging in basic research, drug testing, and drug discovery. Tissue engineering approaches are broadly classified into either scaffold-based or scaffold-free. This chapter focuses on scaffold-free methods and, in particular, the use of micro-molded nonadhesive hydrogels. The use of these hydrogels in biofabrication, data on the formation of a prevascular network in these 3D microtissues, and future directions for multicellular microtissues is discussed.

All scaffold-free methods rely on a phenomenon known as cellular aggregation or *self-assembly*. Mono-dispersed cells seeded in a nonadhesive environment will spontaneously develop cell—cell connections, resulting in cell aggregation. These connections can be formed by a variety of cell adhesion molecules, including cadherins, gap junctions, and tight junctions, depending on the cell type. Scaffold-free biofabrication methods take advantage of this process by creating environments that facilitate cell—cell interactions. It is believed that this increase in cell—cell interactions is responsible for creating tissues that more closely match native tissues in terms of function, protein expression, and mechanical strength. By promoting intrinsic cell—cell connections and communication, the cellular aggregate is able to synthesize extracellular matrix, growth factors, and hormones, so organ function may be replicated on a very small scale. Although the ultimate goal of organogenesis has not yet been achieved and these tissues are small, on the order of hundreds of microns to a few centimeters, scaffold-free methods have been successful in reproducibly creating highly functional microtissues. Most self-assembled aggregates are spherical in shape, called spheroids; some examples include cardiomyocyte spheroids that beat with a human-like rhythm, hepatocyte spheroids with liver-like detoxifying function, chondrocyte microtissues with increased proteoglycan and collagen II content and increased compressive strength, and dorsal root ganglions extending projections through fibroblast spheroids [1—5]. However, a new method developed in our lab is now able to direct the shape of the aggregate into more complex geometries.

During self-assembly, a mixture of two different cell types will often undergo a process known as *self-sorting*, where one cell type forms the inner core of the spheroid and the other cell type forms the outer coating. For many years, it was believed that self-sorting was mediated exclusively by differences in cell adhesion. This theory, the differential adhesion hypothesis (DAH), posited two cell types as immiscible fluids, so one cell type preferentially migrates to the center due to stronger cell surface adhesion interactions compared to the other cell type [6,7]. Numerous studies have supported this hypothesis, but recent work has demonstrated that the process is more complex and also involves cytoskeletal-mediated contraction [8,9].

There are numerous methods for the formation of multicellular spheroids (Table 8.1). *Pellet culture* is a process in which a cell suspension is centrifuged to form a cell pellet wherein the cell—cell interactions result in spheroid formation. The main drawback to this method is that it creates relatively large spheroids whose internal cores may experience low oxygen concentration, with hypoxia leading to cell death [10—13]. *Spinner culture* involves keeping a cell suspension in constant motion using an impeller. Cell—cell collisions result in aggregation, and the constant mixing prevents settling of the cells. The main disadvantages are lack of control over spheroid size and exposure of the cells to shear forces during constant stirring [14—16]. *Rotating wall vessels* filled with a cell suspension are rotated along their *x*-axis, maintaining the cells in a microgravity environment. The rotation is initially slow to allow spheroid formation, and it gradually

Table 8.1 Overview of Methods for Forming Scaffold-Free Multicellular Microtissues

Method	Control Size	Control Shape	Add Soluble Factors	Exposed to Shear Forces	Mass Production
Pellet culture	+	−	+ +	Yes	+
Spinner culture	−	−	+ + +	Yes	+ + +
Hanging drop	+ + +	−	−	No	+
Liquid overlay	−	−	+ + +	No	+
Rotating vessel	−	−	+ + +	Yes	+ + +
External force	+	−	+ + +	Maybe	+
Microfluidics	+ + +	−	+ + +	No	+ +
Micro-molded hydrogels	+ + +	+ + +	+ + +	No	+ + +

increases to higher speed as spheroid size increases. The shear forces are relatively low, but spheroid size is variable [17−19]. *Hanging drops* are created by pipetting small volumes (~20 μL) of a cell suspension onto a flat surface, which is then inverted to allow gravity to collect the cells to the bottom of the drop, where they aggregate and form spheroids. The main drawbacks are the difficulties of extended culture and changes/additions to the culture medium due to small volumes [20−22]. *Liquid overlay* involves seeding a cell suspension onto nonadherent surfaces such as agarose and using gentle shaking to promote cell−cell interaction and aggregation. A drawback is that spheroid size is heterogeneous and difficult to control [23−25]. *External forces* such as electric fields, magnetic fields, or ultrasound are used to concentrate cells within a suspension and induce cell−cell interactions. The main drawbacks are nonspecific cell−cell interactions, difficulty controlling spheroid size, and the uncertain effects of these forces on cell physiology [26−29]. *Microfluidics* involves flowing cell suspensions through micro-channel networks that are partitioned into micro-chambers, where the cells are exposed to micro-rotational flow that induces cell aggregation. These environments can be highly controlled, including spheroid size and the addition of soluble factors, and enable monitoring via biosensors [30−35].

Micro-molded nonadhesive hydrogels, made of agarose or polyacrylamide, are a relatively new method of spheroid formation with a unique capability of forming microtissues with complex shapes beyond the spheroid, thus giving rise to additional applications in biofabrication. Cells are seeded onto a micro-molded nonadhesive hydrogel, and the shape of the small recesses directs the shape of the microtissue, making possible the formation of spheroids, rods, toroids, honeycombs, and other shapes. This chapter discusses this method in depth, presents data related to the self-organization of a prevascular network that occurs in mixed microtissues, and describes how this technology is being used to provide new fundamental information about the self-assembly and self-sorting processes. Finally, the chapter discusses how this technology is being advanced toward goals in biofabrication and the challenges that remain.

8.1 **Materials and methods**

Micro-molds were designed using computer-assisted design (CAD) (SolidWorks Corporation, Concord, MA). Wax molds from the CAD files were produced with a ThermoJet® rapid prototyping machine (3D Systems, Valencia, CA) and replicated in polydimethyl siloxane (PDMS) (Dow Corning, Midland, MI), as previously described [36]. Agarose gels were cast from PDMS micro-molds. Powder UltraPure™ Agarose (Invitrogen, Carlsbad, CA) was sterilized by autoclaving and dissolved via heating in sterile water to 2% (weight/volume). Molten agarose (2.75 mL/mold) was pipetted into each PDMS micro-mold, and air bubbles were removed via pipette suction or agitation with a sterile spatula. After setting, gels were separated from the micro-mold using a spatula, transferred to six-well tissue culture plates, and equilibrated overnight with tissue culture medium. PDMS micro-molds with two different recess geometries were used to produce agarose gels to create spherical or toroidal microtissues. Round recesses to produce spheroids were either 400 or 800 μm in diameter and contained 822 or 330 recesses per gel, respectively. Toroidal recesses were 1400 μm in diameter, with a central agarose peg of 600 μm, and contained 64 features per gel.

Normal human fibroblasts (NHF) derived from neonatal foreskins were expanded in Dulbecco's Modified Eagle Medium (DMEM) (Invitrogen) supplemented with 10% fetal bovine serum (FBS) (Thermo Fisher Scientific, Waltham, MA) and 1% penicillin/streptomycin (Sigma-Aldrich, St. Louis, MO). Human umbilical vein endothelial cells (HUVEC) (Lonza, Basel, Switzerland) were expanded in Endothelial Growth Medium 2 (EGM-2) (Lonza). NHF and HUVEC co-cultures were maintained in a 37°C, 5% CO_2 atmosphere. Cells were trypsinized, counted, and resuspended to the desired cell density for each experiment (Table 8.2). Then 200 μL of a single cell suspension was pipetted into the rectangular seeding chamber above the recesses of each micro-molded agarose gel. Samples were then incubated for approximately 20 minutes to allow cells to settle into recesses before 3 mL of medium was added. DMEM was used for all co-culture experiments. Medium was exchanged every other day.

Brightfield, phase contrast, and fluorescent images were obtained using a Zeiss Axio Observer Z1 equipped with an AxioCam MRm camera with AxioVision Software (Carl Zeiss Micro-Imaging, Thornwood, NY). Fluorescence was provided by the X-Cite 120 fluorescence illumination system (EXFO Photonic Solutions, Ontario, Canada). For observation of immunostaining,

Table 8.2 Experimental Setup of HUVEC-NHF Mixed Microtissues

Approximate		NHF Seeding Number	HUVEC Seeding Number	Agarose Micro-Mold
Diameter	Cells/ Microtissue			
100	120	0.09×10^6	0.01×10^6	822 circular features
300	3000	0.67×10^6	0.33×10^6	330 circular features
500	15,000	3.44×10^6	1.72×10^6	330 circular features
	113,000	5.0×10^6	2.5×10^6	64 toroidal features

microtissues were transferred to glass-bottom fluorodishes (World Precision Instruments, Sarasota, FL) and analyzed with a Leica DM-IRE2 confocal microscope for excitation/detection of 405/475 nm (DAPI), 494/518 nm (FITC), and 595/615 nm (Texas Red). Z-stack images of the lower half of the microtissue were captured in approximately 1-micron intervals. Data are presented as either individual confocal slices or overlays of all slices. Side-view images of microtissues were captured using a Mitutoyo UltraPlan FS110 microscope (Mitutoyo America, Aurora, IL) modified to lie horizontally. To reveal toroidal microtissues, the edge of the gel was cut away using a razor blade. Samples were placed on an adjustable stage, and images were taken with a Nikon Coolpix E990 camera (Nikon USA, Melville, NY) mounted on the eyepiece.

Immunostaining for endothelial cells was performed on microtissues removed from the agarose gel or microtissues within the gel. Microtissues were washed twice in phosphate buffered saline (PBS) and then fixed in 4% paraformaldehyde (Electron Microscopy Sciences, Hatfield, PA) in PBS for 2 hours. Microtissues were washed three times in 0.002% Triton X-100 (Sigma) in PBS, shaking at low pitch for 5 minutes between exchanges and then permeabilized for 2 hours at 4°C in 0.5% Triton X-100 in PBS. Blocking solution, 0.1% Triton X-100, 5% goat serum (Sigma), and 1% bovine serum albumin (BSA) (Sigma) in PBS were added for 2 hours. Next, primary antibody (20 μg/mL) was added in 5% goat serum with 1% BSA in PBS, and incubated for at least 8 hours at 4°C. Primary antibodies were a mouse monoclonal antibody to human CD31 and a rabbit polyclonal antibody to human von Willebrand factor (vWF) (Sigma). Microtissues were washed in blocking solution for 2 hours and then three times in PBS, shaking at low pitch for 5 minutes between exchanges. Secondary antibodies were added at a 1:100 dilution in 5% goat serum with 1% BSA in PBS and incubated for at least 8 hours. Secondary antibodies were Texas Red-X goat antimouse IgG and FITC goat antirabbit IgG (20 ng/mL)

(Sigma). DAPI (5 ng/mL) (Invitrogen) was used as a counterstain. After incubation, microtissues were washed three times in PBS and observed by confocal or widefield epifluorescent microscopy.

8.2 Results

When a mono-dispersed cell suspension was pipetted over an array of recesses micro-molded into agarose, the cells settled into the recesses. Since the cells are unable to attach to the agarose, the small forces of cell aggregation drive the self-assembly of a spheroid, one per recess (Figure 8.1). NHF and HUVEC cells seeded at a 2:1 ratio resulted in 300-μm-diameter spheroids that self-sorted to have an NHF core coated by HUVECs. After 1 week of culture, spheroids were stained with a monoclonal antibody to CD31, a marker of endothelial differentiation. HUVECs displayed elongated, overlapping morphologies on the inside of the spheroids, suggesting the formation (self-organization) of a cellular prevascular network.

To determine the effect of spheroid size and culture time on the formation of a prevascular network, spheroids with diameters of approximately 100, 300, and 500 μm cultured for 1, 4, and 7 days, respectively, were immunostained (Figure 8.2). After 1 day, no CD31-positive cells were observed in any of the experimental groups. Additionally, no CD31-positive cells were seen in the 100-μm-diameter spheroids at any experimental endpoint. A CD31-positive

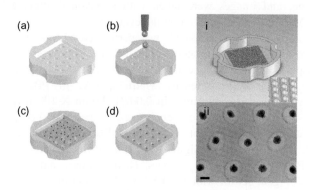

FIGURE 8.1

Micro-molded nonadhesive hydrogel to form spheroids (a). Mono-dispersed cells were pipetted into the square seeding chamber over the array of recesses (b, c). Gravitational forces cause cells to sink into micro-molded recesses, leading to the formation of microtissues (d). (i) A CAD image of the negative mold used to form the micro-molded agarose gel. Inset: Detail of the pegs that lead to round-bottom wells. (ii) An image of spheroids formed within the recesses of micro-molded agarose gel.

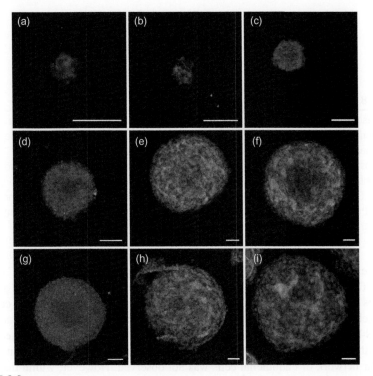

FIGURE 8.2

The formation of an endothelial prevascular network is dependent on spheroid size and time in culture. NHFs and HUVECs were seeded at a 2:1 ratio such that the resulting spheroids were 100 (a, b, c), 300 (d, e, f), or 500 (g, h, i) microns in diameter. Cells were fixed and stained with DAPI (blue) and a monoclonal antibody to CD31 (red) at 1 (a, d, g), 4 (b, e, h), and 7 (c, f, i) days. A prevascular network was seen in spheroids larger than 300 microns after four days. Images are overlays of confocal microscopy. Scale bars are 100 μm.

prevascular network was observed in spheroids with a diameter larger than 300 microns after 4 days of culture.

These data suggest that the formation of a prevascular network is both time- and spheroid size-dependent. During the first 24 hours, spheroids self-assembled, and the NHFs and HUVECs self-sorted to form the inner core and outer coating, respectively. Over the next few days, the endothelial cells invaded the NHF core and differentiated to form the prevascular network. This demonstrates that after the initial self-assembly and self-sorting are complete, the spheroid is not necessarily a static structure. Sufficiently large spheroids are a dynamic cellular environment that not only produce signals for endothelial cell migration and differentiation but also are capable of remodeling.

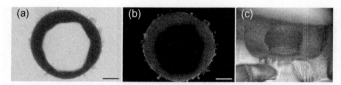

FIGURE 8.3

A prevascular network is not formed in toroidal microtissues. Brightfield image of a toroidal microtissue generated from over 100,000 cells wrapped around an agarose peg has a mean minor axis of 126 μm (a). Fluorescent image of a toroidal microtissue immunostained for CD31 and vWF with DAPI counterstain (b). A prevascular network of endothelial cells was not observed, forming only a thin layer of endothelial cells along the outside. Side-view brightfield image of a toroidal microtissue assembled upward and around the agarose peg has a collar-like structure with a mean thickness of 407 μm (c). Scale bars are 200 μm.

To examine the effects of microtissue geometry, NHFs and HUVECs were mixed (2:1) and seeded into an agarose micro-mold containing toroid-shaped recesses (Figure 8.3). Samples were fixed and stained after 5 days of culture. Cells self-assembled into a toroid-shaped microtissue with mean minor axis of 126 ± 44.5 μm (standard deviation, n = 20). Endothelial cells self-sorted and coated the outside of the microtissue as labeled by an antibody to vWF, but a prevascular network of endothelial cells was not observed. However, within the same experiment, several toroidal microtissues released from the micro-mold and collapsed into a spheroidal microtissue. Immunostaining revealed that HUVECs coated the outside of these microtissues and stained positively with von Willebrand factor (vWF), while a CD31-positive vascular-like network within the resulting large-sized spheroid was also observed, consistent with previous findings. The geometry of toroidal microtissues was investigated further using side-view microscopy. This side-view image revealed that the microtissue self-assembled upward around the peg in a collar-like structure with a mean tissue thickness of 407 ± 38 μm (n = 14).

It is hypothesized that decreased oxygen levels at the center of large (>300 μm) microtissues results in a local hypoxic response that stimulates the angiogenic response in spheroids. Interestingly, the self-organization of the prevascular network did not occur in toroidal microtissues, even though these microtissues had far more cells (113,000 vs. 15,000) per microtissue than the large 500-μm-diameter spheroids. Unlike a spherical geometry where the distance to the surface (radius) is the same in all three dimensions, the distance of the innermost cell to the surface of a toroidal microtissue is not the same in all three dimensions. The plane about which the toroidal microtissue is rotated never contacts the surface. The thickness of our toroidal microtissues was 400 μm, a distance expected to induce the angiogenic response, but the minor axis of the toroid was only 126 μm, a distance not expected to induce the angiogenic response.

These data suggest that the hypoxic response might not be induced if the distance to the surface in one of three possible dimensions is less than $100-200\,\mu m$, implying that microtissue geometry may influence prevascular formation.

8.3 **Discussion**

All of the methods for forming scaffold-free multicellular microtissues harness the same biological phenomenon of cell−cell aggregation (self-assembly) and the related self-organization (self-sorting) that can occur as a mixture of different cell types self-assembles. An increasingly long list of cell types, including primary cells, cell lines, stem cells, and tumor cells, are competent to self-assemble, suggesting that this is an intrinsic property of nearly all adherent cell types. Self-assembly and self-sorting are complex biological processes involving numerous types of surface adhesion molecules working in concert with the proteins of the cytoskeleton. Although the cellular mechanisms controlling self-assembly are not fully understood, there are numerous methods to produce functional 3D microtissues with applications in biofabrication. Thus, there are two major objectives that are synergistic. The first is to increase the fundamental understanding of the complex process of self-assembly so that it can be better controlled. The second is to expand the range of use of scaffold-free multicellular microtissues in tissue engineering and biofabrication. We are using one of the methods, micro-molded nonadhesive hydrogels, to form microtissues and to address both objectives, and this chapter reviews some of that progress as well as some of the challenges that remain.

Unlike other methods that form only spheroids, micro-molded nonadhesive hydrogels (micro-molds) can form microtissues with complex shapes. Since micro-mold design is controlled by computer-aided design (CAD), these can be simple shapes such as circular recesses in which the cells form spheroids, or they can be complex shapes such as large ($\sim 2\,cm$) honeycomb structures [36,37]. When cells are seeded onto micro-molds with complex shapes, they settle by gravity onto the micro-mold and undergo an initial adhesion with neighboring cells. Subsequently, this loosely adhered group of cells begins the more complex process of self-assembly, which results in significant morphological changes to this collection of cells. As self-assembly proceeds, cells form increasing numbers of cell−cell adhesions and also exert cell−cell contractile forces that are transmitted through these adhesions. These collective forces of self-assembly drive the 3D morphological changes that occur over time. Micro-mold design serves two purposes: It sets the initial starting configuration of the cells (e.g., shape of the footprint of the settled cells), and it provides nonadhesive obstacles (e.g., posts) whose size, shape, and position direct self-assembly, thereby controlling the morphology of the microtissue as the cells self-assemble.

The hydrogel is easily micro-molded and has several inherent properties important to its usefulness in this application. First, the hydrogel is nonadhesive for cells. In stark contrast to a standard tissue culture plate that promotes vigorous cell attachment and cell spreading, cells are unable to attach to the nonadhesive agarose or polyacrylamide hydrogel. This allows the relatively small forces of cell—cell aggregation to drive self-assembly, which is typically complete within 24 hours. Second, the hydrogel environment ensures that microtissues are uniformly surrounded by culture medium on all surfaces. There are no unnatural barriers to nutrient diffusion or artificial gradients imposed by solid substrates. Moreover, the stable hydrogel environment facilitates long-term culture (at least 2 weeks), easy replenishments of culture medium, and serial addition of drugs, growth factors, and so on. Finally, since the microtissues are not attached to the hydrogel, they can be easily harvested by inverting the gel and then can be used in many applications, including building blocks for biofabrication.

Since the micro-molds are made using CAD and rapid prototyping, there are unlimited design possibilities, and some of these designs have been used as tools to investigate the fundamentals of the self-assembly process. A *trough design*, in which cells form a multicellular rod-shaped microtissue whose long axis contracts with time, has been used to quantify the rate of self-assembly. Using this assay, it was demonstrated that different cell types vary significantly in their rates and extents of self-assembly. Rods of NHFs contracted with an exponential rate within 12 hours, whereas rods of a hepatocyte cell line (H35) contracted with linear kinetics over several days. NHF rods contracted to a final spheroid structure, whereas H35 rods contracted, but still remained in a rod structure with a long axis. Interestingly, when the length of the trough was systemically increased, H35 rods were longer, but the rod was consistently 49 percent of trough length for all lengths tested. This consistency in self-assembly with respect to micro-mold design (e.g., percent of trough length) suggests that given a specific micro-mold design, the final size and shape of a self-assembled microtissue may be accurately predicted [38]. The rate of rod contraction has also been used to investigate the role of specific proteins in self-assembly. Using this assay, it has been shown that the membrane proteins connexon 43 (gap junction) and pannexin-1 are important to self-assembly [39,40].

A *loop-ended dogbone design* has been used to investigate the effects of obstacles on self-assembly. In this design, two circular pegs are surrounded by a continuous trough that extends to form the linear trough between the two obstacles. The multicellular microtissue formed by this design is two toroids connected by a rod. Unlike the rod formed in a simple trough, the loop-ended dogbone is constrained by the pegs. These constraints elicit uniaxial tension in the connecting rod as self-assembly proceeds. In the case of NHFs, which exert significant forces, the connecting rod thins over time (also called necking) and ultimately fails. Individual cells within the microtissue actively increase their surface area during self-assembly by extending and retracting cell projections. Prior to microtissue failure, cells align and elongate up to 30 times their initial diameter.

Drugs that target the cytoskeleton and its motors (Y-27632 and blebbistatin) reduce this tension and prevent failure. Cell types that do not exert as much force as NHFs do not fail in this design [41]. For the purposes of biofabrication, these data show that cell types differ with regard to their ability to form stable complex-shaped microtissues as directed by obstacles and that stability is related to the level of cytoskeletal-mediated contraction.

A *trough around a cone design* has been used to precisely quantify the work performed by cells as they self-assemble. Cells form a multicellular toroid, and the forces of self-assembly drive this toroid up the nonadhesive cone against the force of gravity. The rate at which these cellular forces move the toroid up the cone is power: $P = \Delta W/\Delta t$, where ΔW is work necessary to move a toroid of a given mass to a given height, and Δt is the time over which the work is performed. Thus, power is the quantification of the rate of self-assembly in terms of the work output. The power of an NHF toroid is 4.3 ± 1.7 pJ/hr and of an H35 toroid it is 0.31 ± 0.01 pJ/hr, a major difference between two different cell types. This assay has been used to quantitatively dissect the contributions of specific parts of the multicomponent system that drives self-assembly. Blocking *rho* kinase with the drug Y-27632 resulted in a 50 percent or greater reduction in power expended by both NHF and H35 toroids. The assay was also used to quantify the work performed by a mixture of cells. The power of a small number of NHFs ($\sim$10 percent) was significantly enhanced when they were mixed with H35 cells [42]. For purposes of biofabrication, these data provide a precise quantitative measure of self-assembly and how blends of different cell types interact when they self-assemble.

Micro-mold designs have also been used to investigate the phenomenon of self-sorting. A *circular well design* for forming spheroids was used to show that in addition to differences in surface adhesion, differences in cytoskeletal-mediated tension influence self-sorting. NHFs treated with Y-27632 sorted to the outside of the spheroid when co-seeded with untreated NHFs [9]. Other studies with spheroids demonstrated that cell organization is not fixed after self-assembly and that there is a level of fluidity in a self-assembled spheroid. A mixture of NHFs and human umbilical vein endothelial cells (HUVECs) will self-sort to form the core and coating, respectively. If a HUVEC spheroid is formed and matured for 7 days, newly added mono-dispersed NHFs are still able to sort to the core [8]. The data presented in this chapter (Figures 8.2 and 8.3) used spheroids and toroids to show that self-sorting and the formation of a prevascular network by a mix of NHFs and HUVECs are dependent both on time and on the geometry of the microtissue. In another study using the toroid design, it was shown that TGF-β1 treatment alters self-sorting [43]. Spheroids were also used to develop a fluorescent image-based method to quantify the rate and extent of self-sorting. The rate of self-sorting of NHFs and H35s was identical to the overall rate of self-assembly, even when the proportion of NHFs was varied over a sixfold range (14 to 85 percent). This quantitative tool can characterize the self-sorting of any pair of cells, providing new information on the cellular mechanisms of self-sorting and helping to identify drugs or growth factors that can be used to control self-sorting [44]. For purposes of biofabrication, the ability to control self-sorting will make possible the formation of designer microtissues where the position of different cell types is specified.

Micro-molds have been used to produce building blocks for applications in biofabrication. As shown in the data presented in this chapter, a single pipetting step creates hundreds of spheroids, so the scaling up of building block production is relatively straightforward. The size of the spheroids is uniform, and because cells settle from the seeding chamber into the individual recesses, spheroid size is controlled by the total number of cells seeded onto the micro-mold. Spheroids as small as 25 cells to hundreds of cells per spheroid have been produced. As shown in the data reported in this chapter, as well as that published elsewhere, mixed spheroids with two or even three different cell types can be produced. The ratio of each cell type per spheroid is controlled by adjusting the ratio of cell types in the mixture of mono-dispersed cells pipetted onto the micro-mold [8].

In addition to spheroids, micro-molds have been used to make building parts with more complex shapes, including rods, toroids, and honeycombs. As with spheroids, it is easy to scale up the production of microtissues with these shapes, and these shapes have been made with mixtures of different cell types. Even in complex shapes, self-sorting occurs as the microtissue self-assembles. When harvested from the micro-molds, microtissues with complex shapes remain intact. These nonspheroidal shapes provide new possibilities for building parts, and the simple toroid, with its ring of cells in high density and open lumen space, best exemplifies this potential. Unlike the spheroid shape, where diffusion limits its maximum size (<200 μm diameter), toroid building units can be made over a range of sizes without compromising cell viability, provided that *one* of the tissue dimensions does not exceed the diffusion limit [45]. The viability of the toroid stems from its open lumen with access to the medium. While the outermost diameter of the entire toroid is large, the cross section of the cellular portion of the toroid does not exceed the critical diffusion distance needed to maintain cell viability. If the same number of cells were self-assembled into a single spheroid, its diameter would exceed the critical distance needed to maintain the viability of the cells in the core of the spheroid [37,46].

Using changes to micro-mold design, toroids with lumens that ranged in diameter from 1000 μm down to 400 μm were produced. Lumen size was controlled by the diameter of the agarose peg in the micro-mold, and 300−400 μm was the smallest diameter that could reliably be made using rapid prototyping technology. Other micro-fabrication technologies with higher resolution could potentially produce micro-molds with smaller pegs to create toroids with even smaller lumens. The morphological stability of a building part after it is harvested from a micro-mold is an important property, and as a living structure this morphology is not necessarily static. When harvested, the toroids remained intact, undergoing predictable changes to their size and shape, but retained a patent lumen even after ten days [46]. The observation that these morphological changes were predictable means that ultimately they can be factored into the design of the building part.

Another variation on the toroid building unit is the honeycomb, a larger structure with multiple lumens. Micro-molds have been used to produce honeycombs as large as approximately 2 cm (7×10^6 cells) containing 217 lumens organized into eight orbitals around a central lumen. As a building unit, the honeycomb is significantly larger, with far more cells than a spheroid or a toroid, yet it can be self-assembled in approximately the same time (approximately 24 hours). Like the toroid, the honeycomb has open lumens with access to cell culture medium. While the overall x and y dimensions of a honeycomb can be quite large, the cross section of the cellular portion of any part of the honeycomb can be kept small so as not to exceed the critical diffusion distance needed to maintain cell viability. In addition to size, another advantage over the toroid is that the honeycomb fixes the relative positions of many lumens in the x-y plane, thus providing more premade organization as a building part. Unlike the spheroid, whose size is constrained by a critical diffusion distance needed to maintain cell viability, the overall size of the honeycomb in the x-y plane is not limited, so honeycomb building parts with large surface areas can be formed [37].

Critical to the usefulness of any building part in biofabrication is its ability to be used to build larger structures. When contacted with one another, building parts will undergo a process of *microtissue fusion*, where adjacent parts meld to form a single continuous microtissue. Fusion is complete in about 48 hours. Micro-molds have been used to develop a quantitative assay to investigate this biological process of microtissue fusion. Harvested spheroids were placed in a micro-mold with a trough design where adjacent spheroids melded to form a rod whose rate of contraction was measured over time. Increasing the time of maturation of NHF spheroids (1,4, and 7 days after self-assembly) significantly slowed the rate at which they fused, and they formed longer rods. The fusion kinetics and steady state length of rods formed by smaller versus larger spheroids (~100 versus 300 μm diameter) were indistinguishable, even though smaller spheroids had twice the surface area and greater numbers of contacts between building units. Like large spheroids, small spheroids were strongly influenced by spheroid maturation time. The work was extended to mixtures of different cell types, and spheroid maturation was also able to control the 3D position of cells in a melded microtissue. This study also showed that micro-molds could be used to guide this melding process to form complex shapes. Spheroid building blocks were added to secondary micro-molds (toroid or honeycomb), where they fused to form these shapes [47].

Building parts with complex shapes have been melded to form even larger structures. Toroids, self-assembled from mono-dispersed cells, fuse within 48 to 72 hours. When the outer edge of a pair of adjacent toroids was contacted, melding occurred, and when a pair of toroids were stacked on a nonadhesive hydrogel cone, their top and bottom surfaces fused. The use of a cone is another example of a micro-mold being used to guide the melding process. Fusing of fluorescently labeled toroids showed that minimal cell mixing occurs between the building units. A large number of toroids were fused to form a multitoroid structure. Toroids added to a single well settled to form a pile of toroids that fused over time. The bias in the way that toroids settled resulted in a large, open structure with interconnected lumens where toroids overlapped, but the majority of toroids had their lumens oriented along the z-axis [46].

8.4 Future considerations

A major challenge to the field of biofabrication is the in vitro fabrication of large, solid organs with high densities of living cells. Diffusion of oxygen and nutrients and the removal of metabolic waste products limit current engineered tissues to thicknesses of approximately $100-200\,\mu m$ in order to maintain cell viability. Natural organs and tissues are much larger and contain a branching vascular network that perfuses the entire organ and ensures all cells are close to blood vessels with adequate nutrient and oxygen supply. As the field of biofabrication and tissue engineering struggles with this major limitation, the field of induced pluripotent (iPS) stem cells is well on its way to providing a plentiful source of immune-matched cells of a variety of tissues and organs. Although we do not yet have a means for the in vitro fabrication of large 3D organs and tissues from this source of cells, multicellular building blocks may someday be useful for this problem.

In their present form, microtissue building blocks are not able to fabricate large vascularized organs in vitro, but they do approximate the size and complexity of pancreatic islets, with islet diameters of approximately 200 microns. Given the correct source of differentiated cells, technologies such as micro-molds that can be scaled up would be able to produce sufficient numbers of islets for transplantation to achieve insulin independence (approximately 10,000 islets per kg body weight). With regard to the effort to fabricate large vascularized organs in vitro, building blocks may be helpful if they are combined with other technologies used in biofabrication. Bioprinting, a well-advanced technology described in another chapter in this book, is based on the principle of inkjet printing and uses cells and ECM materials to build 3D constructs layer by layer [48,49]. Building blocks produced using micro-molds may be useful as a material for bioprinting. Cell sheets are another technology that might be combined with microtissue building blocks. Cell sheets are produced by culturing cells attached to a thermo-responsive polymer. When the cells reach confluence, the temperature is decreased, causing the release of an intact cell sheet [50,51]. Building blocks of defined geometries might be combined with cell sheets to form larger structures. Finally, building blocks might be combined with scaffolds, a large and very active area of investigation. Building blocks will readily attach to and spread on the many natural and synthetic scaffolds that have been engineered for cell attachment.

If the goal is to use solely building blocks to fabricate a large vascularized organ in vitro, there are significant challenges yet to be overcome, and it is informative to discuss these challenges in the context of the current progress in that direction. Regardless of their shape, building blocks are rapidly self-assembled from mono-dispersed cells within 24 to 48 hours. Micro-molds can be scaled up to produce sufficient numbers of building blocks, and building blocks can be made as very large structures (e.g., honeycombs) so fewer parts are needed for construction. These building parts can be melded within 48 and 72 hours, so the assembly of a large construct from individual parts could theoretically be accomplished in a relatively short period of time (1 to 2 weeks). This, of course, depends on the time required for maturation and does not count the time necessary to culture the large number of cells needed to form building parts. Building parts with lumens (toroids) have begun to approximate a crude vascular network, and the fusion of overlapping toroids generates a size range of lumens, all smaller than the lumen of the building unit. However, these neo-lumens approximate neither the density nor the diameter of capillaries ($\sim$10 μm). They may be useful for recreating the range of vessels that connect capillaries to small-diameter arteries and veins ($\sim$0.1—5.0 mm).

It may be possible to form building blocks that have a preformed capillary network. As presented here, endothelial cells will self-organize and form a pre-vascular network as a mixed microtissue self-assembles. Also needed are strategies to endothelialize the lumens of the building blocks to determine if these lumens anastomose with the capillary network and if the bioengineered vascular tree can be perfused as it is being built and as it matures. Since the critical diffusion distance for cells to receive necessary nutrients and to offload waste products is approximately 150 microns, perfusion is critical to maintain cell viability in large bioengineered structures with a high density of cells. A robust and functional vascular supply is also critical from the surgical perspective if a biofabricated tissue is to be translated to the operating room. A tissue with high cell density will require integration into the blood supply through arterial and venous anastomosis with the host. For tissues that are body site—specific, such as a renal tissue that requires a urinary drainage system or a hepatic tissue that requires a biliary drainage system, the scale of the engineered tissue must be adequate to properly and safely anastomose the drainage portion to the corresponding host organ. In summary, tissue engineering, and specifically scaffold-free tissue engineering, has come a long way, but there is still much progress to be made, and keeping the ultimate application of that work in mind should guide methods and designs of engineered tissues. These are just a few of the many challenges and clearly much work needs to be done to make advances toward this goal.

References

[1] Kelm JM, Fussenegger M. Microscale tissue engineering using gravity-enforced cell assembly. Trends Biotechnol 2004;22(4):195–202.

[2] Fukuda J, Nakazawa K. Orderly arrangement of hepatocyte spheroids on a microfabricated chip. Tissue Eng 2005;11(7–8):1254–62.

[3] Fukuda J, Sakai Y, Nakazawa K. Novel hepatocyte culture system developed using microfabrication and collagen/polyethylene glycol microcontact printing. Biomaterials 2006;27(7):1061–70.

[4] Ofek G, et al. Matrix development in self-assembly of articular cartilage. PLoS One 2008;3(7):e2795.

[5] Kelm JM, et al. Self-assembly of sensory neurons into ganglia-like microtissues. J Biotechnol 2006;121(1):86–101.

[6] Steinberg MS. Mechanism of tissue reconstruction by dissociated cells. II. Time-course of events. Science 1962;137:762–3.

[7] Foty RA, et al. Surface tensions of embryonic tissues predict their mutual envelopment behavior. Development 1996;122(5):1611–20.

[8] Napolitano AP, et al. Dynamics of the self-assembly of complex cellular aggregates on micromolded nonadhesive hydrogels. Tissue Eng 2007;13(8):2087–94.

[9] Dean DM, Morgan JR. Cytoskeletal-mediated tension modulates the directed self-assembly of microtissues. Tissue Eng Part A 2008;14(12):1989–97.

[10] Jahn K, et al. Pellet culture model for human primary osteoblasts. Eur Cell Mater 2010;20:149–61.

[11] Giovannini S, et al. Micromass co-culture of human articular chondrocytes and human bone marrow mesenchymal stem cells to investigate stable neocartilage tissue formation in vitro. Eur Cell Mater 2010;20:245–59.

[12] Li J, He F, Pei M. Creation of an in vitro microenvironment to enhance human fetal synovium-derived stem cell chondrogenesis. Cell Tissue Res 2011;345(3):357–65.

[13] Markway BD, et al. Enhanced chondrogenic differentiation of human bone marrow-derived mesenchymal stem cells in low oxygen environment micropellet cultures. Cell Transplant 2010;19(1):29–42.

[14] Nyberg SL, et al. Rapid, large-scale formation of porcine hepatocyte spheroids in a novel spheroid reservoir bioartificial liver. Liver Transpl 2005;11(8):901–10.

[15] Frith JE, Thomson B, Genever PG. Dynamic three-dimensional culture methods enhance mesenchymal stem cell properties and increase therapeutic potential. Tissue Eng Part C Methods 2010;16(4):735–49.

[16] Han Y, et al. Cultivation of recombinant Chinese hamster ovary cells grown as suspended aggregates in stirred vessels. J Biosci Bioeng 2006;102(5):430–5.

[17] Ingram M, et al. Three-dimensional growth patterns of various human tumor cell lines in simulated microgravity of a NASA bioreactor. In Vitro Cell Dev Biol Anim 1997;33(6):459–66.

[18] Carpenedo RL, Sargent CY, McDevitt TC. Rotary suspension culture enhances the efficiency, yield, and homogeneity of embryoid body differentiation. Stem Cells 2007;25(9):2224–34.

[19] Manley P, Lelkes PI. A novel real-time system to monitor cell aggregation and trajectories in rotating wall vessel bioreactors. J Biotechnol 2006;125(3):416–24.

[20] Kelm JM, et al. Method for generation of homogeneous multicellular tumor spheroids applicable to a wide variety of cell types. Biotechnol Bioeng 2003;83(2): 173−80.

[21] Timmins NE, et al. Method for the generation and cultivation of functional three-dimensional mammary constructs without exogenous extracellular matrix. Cell Tissue Res 2005;320(1):207−10.

[22] Tung YC, et al. High-throughput 3D spheroid culture and drug testing using a 384 hanging drop array. Analyst 2011;136(3):473−8.

[23] Enmon Jr. RM, et al. Dynamics of spheroid self-assembly in liquid-overlay culture of DU 145 human prostate cancer cells. Biotechnol Bioeng 2001;72(6):579−91.

[24] Metzger W, et al. The liquid overlay technique is the key to formation of co-culture spheroids consisting of primary osteoblasts, fibroblasts and endothelial cells. Cytotherapy 2011;13(8):1000−12.

[25] Landry J, et al. Spheroidal aggregate culture of rat liver cells: histotypic reorganization, biomatrix deposition, and maintenance of functional activities. J Cell Biol 1985;101(3):914−23.

[26] Sebastian A, Buckle AM, Markx GH. Tissue engineering with electric fields: immobilization of mammalian cells in multilayer aggregates using dielectrophoresis. Biotechnol Bioeng 2007;98(3):694−700.

[27] Ino K, Okochi M, Honda H. Application of magnetic force-based cell patterning for controlling cell-cell interactions in angiogenesis. Biotechnol Bioeng 2009;102(3): 882−90.

[28] Okochi M, et al. Three-dimensional cell culture array using magnetic force-based cell patterning for analysis of invasive capacity of BALB/3T3/v-src. Lab Chip 2009;9(23):3378−84.

[29] Liu J, et al. Functional three-dimensional HepG2 aggregate cultures generated from an ultrasound trap: comparison with HepG2 spheroids. J Cell Biochem 2007;102(5): 1180−9.

[30] Okuyama T, et al. Preparation of arrays of cell spheroids and spheroid-monolayer cocultures within a microfluidic device. J Biosci Bioeng 2010;110(5):572−6.

[31] Huang CP, et al. Engineering microscale cellular niches for three-dimensional multicellular co-cultures. Lab Chip 2009;9(12):1740−8.

[32] Hsiao AY, et al. Microfluidic system for formation of PC-3 prostate cancer co-culture spheroids. Biomaterials 2009;30(16):3020−7.

[33] Jin HJ, et al. A multicellular spheroid formation and extraction chip using removable cell trapping barriers. Lab Chip 2011;11(1):115−9.

[34] Agastin S, et al. Continuously perfused microbubble array for 3D tumor spheroid model. Biomicrofluidics 2011;5(2):24110.

[35] Toh YC, et al. A novel 3D mammalian cell perfusion-culture system in microfluidic channels. Lab Chip 2007;7(3):302−9.

[36] Napolitano AP, et al. Scaffold-free three-dimensional cell culture utilizing micromolded nonadhesive hydrogels. Biotechniques 2007;43(4):494, 496−500.

[37] Tejavibulya N, et al. Directed self-assembly of large scaffold-free multi-cellular honeycomb structures. Biofabrication 2011;3(3):034110.

[38] Dean DM, et al. Rods, tori, and honeycombs: the directed self-assembly of microtissues with prescribed microscale geometries. FASEB J 2007;21(14):4005−12.

[39] Bao B, et al. Connexon-mediated cell adhesion drives microtissue self-assembly. FASEB J 2011;25(1):255−64.

[40] Bao BA, et al. Pannexin1 drives multicellular aggregate compaction via a signaling cascade that remodels the actin cytoskeleton. J Biol Chem 2012;287(11):8407−16.

[41] Dean DM, Rago AP, Morgan JR. Fibroblast elongation and dendritic extensions in constrained versus unconstrained microtissues. Cell Motil Cytoskeleton 2009;66(3):129−41.

[42] Youssef J, et al. Quantification of the forces driving self-assembly of three-dimensional microtissues. Proc Natl Acad Sci USA 2011;108(17):6993−8.

[43] Youssef J, et al. Mechanotransduction is enhanced by the synergistic action of heterotypic cell interactions and TGF-beta1. FASEB J 2012;26(6):2522−30.

[44] Achilli TM, et al. Quantification of the kinetics and extent of self-sorting in three dimensional spheroids. Tissue Eng Part C Methods 2012;18(4):302−9.

[45] Colton CK. Implantable biohybrid artificial organs. Cell Transplant 1995;4(4):415−36.

[46] Livoti CM, Morgan JR. Self-assembly and tissue fusion of toroid-shaped minimal building units. Tissue Eng Part A 2010;16(6):2051−61.

[47] Rago AP, Dean DM, Morgan JR. Controlling cell position in complex heterotypic 3D microtissues by tissue fusion. Biotechnol Bioeng 2009;102(4):1231−41.

[48] Wilson Jr. WC, Boland T. Cell and organ printing 1: protein and cell printers. Anat Rec A Discov Mol Cell Evol Biol 2003;272(2):491−6.

[49] Jakab K, et al. Tissue engineering by self-assembly and bio-printing of living cells. Biofabrication 2010;2(2):022001.

[50] Shimizu T, et al. Fabrication of pulsatile cardiac tissue grafts using a novel 3-dimensional cell sheet manipulation technique and temperature-responsive cell culture surfaces. Circ Res 2002;90(3):e40.

[51] Sekine H, et al. Cardiac cell sheet transplantation improves damaged heart function via superior cell survival in comparison with dissociated cell injection. Tissue Eng Part A 2011;17(23−24):2973−80.

A Digital Microfabrication-Based System for the Fabrication of Cancerous Tissue Models

9

Qudus Hamid,[1] Chengyang Wang,[1] Jessica Snyder,[1] Yu Zhao,[2] Wei Sun,[1,2,3]

[1]*Department of Mechanical Engineering and Mechanics, Drexel University, Philadelphia, PA, USA;* [2]*Mechanical Engineering and Biomanufacturing Research Institute, Tsinghua University, Beijing, China;* [3]*Shenzhen Biomanufacturing Engineering Laboratory, Shenzhen, Guangdong, China*

CONTENTS

INTRODUCTION

Micro-electro-mechanical systems (MEMS) technologies have been very attractive and have demonstrated the potential for many applications in the field of

Biofabrication.

tissue engineering, regenerative medicine, and life sciences. These fields bring together the multidisciplinary field of engineering and integrated sciences to fabricate tissue models that aid the exploration, generation, or regeneration of organic tissues and organs [1−3]. MEMS were first introduced on conventional semiconductor materials, which were used to develop integrated circuits (ICs). Since the development of the first IC with MEMS technologies, MEMS have been utilized in many other fields with great success. MEMS technology, however, is expensive and has many limitations [4−6].

There is an overwhelming need for substitutes to repair tissues and organs because of disease, trauma, or congenital problems. In the United States alone, as many as 20 million patients per year suffer from various organ- and tissue-related maladies caused by burns, skin ulcers, diabetes, and connective tissue defects, which include bone and cartilage damage. More than 8 million surgical procedures are performed annually to treat these cases, over 70,000 people are on transplant waiting lists, and an additional 100,000 patients die due to the lack of availability of appropriate organs [7]. Scientists are working around the clock to develop pharmaceuticals and tissue replacements that would allow humans to live longer lives. However, many of these developments require tremendous amounts of research on their effects on humans. Quite often, the use of animal and human models is limited by the feasibility of testing protocols, availability, and ethical anxieties [8,9]. The digital micro-mirroring microfabrication (DMM) system will give scientists the capabilities to develop models that can be utilized to characterize new pharmaceuticals and tissue replacements and to develop models to study fatal diseases such as cancers [10−12]. This biologically inspired microfabrication system has the potential to develop critical three-dimensional models for the investigation of various tissue models and biological sensors. Three-dimensional biological models are preferred for in vitro investigation, since these models eliminate the limitations of traditional mainstay two-dimensional models [13,14].

The DMM system can be used to fabricate many advantageous devices. Among them, microfluidics systems have the most tissue engineering, regenerative medicine, and life sciences applications to develop in vitro tissue models. Unlike many conventional microfabrication techniques and/or devices available to fabricate tissue models, this system eliminates the need for masks by incorporating a dynamic maskless fabrication technique [15−18]. Since the DMM system can develop models on a micro-scale level, this makes research more economic, requiring fewer reagents and fewer cells, and it will allow for consistency in experimental analysis to perform limited interactions with end users [19−21]. The DMM system is specifically designed for the development of biologically inspired devices, which include, but are not limited to, biosensors, Lindenmayer systems, and micro-organs (Figure 9.1). This fabrication system eliminates the limitations of conventional photolithography and gives the end user the ability to develop advantageous models [16,22].

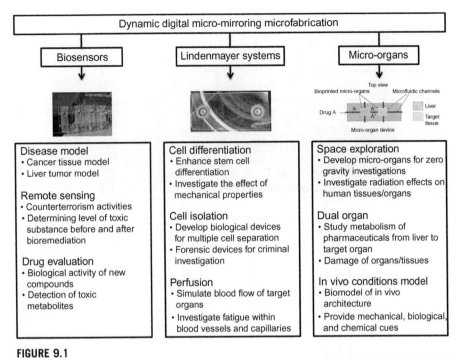

FIGURE 9.1

Applications of the dynamic digital micro-mirroring microfabrication system.

9.1 Digital micro-mirroring microfabrication system

The DMM system consists of three major components: a digital micro-mirroring projection system, a photolithographic substrate alignment system, and a mask modeling system. The micro-mirroring projection system is connected to a computer interface. The computer system activates and deactivates the projection of the mask onto the substrate. The photolithographic substrate alignment system consists of a digital microscopic device that allows for alignment of the substrate's features. The mask modeling system utilizes computer-aided design (CAD) technologies to design a mask for projection. The mask projected by the micro-mirrors must be in .jpeg, .bitmap, or .gif format. Figure 9.2 illustrates the digital micro-mirroring microfabrication system and its corresponding flow diagram.

9.1.1 Digital micro-mirroring projection system

The micro-mirroring projection system has the potential to switch between masks within a matter of microseconds, while offering high-resolution performance in spatial light modulation (SLM). With ultraviolet (UV) light, this system offers a

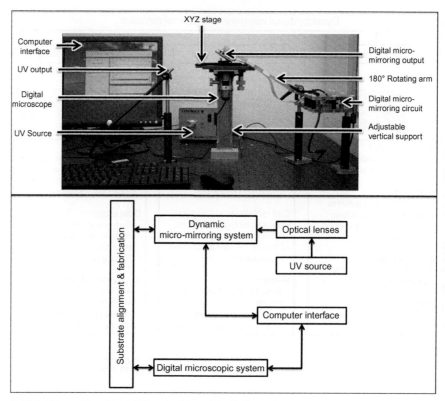

FIGURE 9.2

(Top) Digital micro-mirroring microfabrication system. (Bottom) Structure of the digital micro-mirroring microfabrication system.

flexible platform to design and develop proof of concepts, tissue models, biosensors, micro-organs, and Lindenmayer systems. The main component of the digital micro-mirroring projection system is the digital micro-mirror device (DMD), which is an optical semiconductor module that allows the digital manipulation and projection of UV light. The DMD is mounted directly above the alignment platform and is angled toward the UV light source. The DMD is completely adjustable in terms of elevation and angle. The UV light source emits an adjustable light in terms of intensity and exposure time. The light from the UV lamp travels and uniformly distributes on the micro-mirrors. Figure 9.3 shows the control system diagram of the digital micro-mirroring projection system and the direction of the light projected onto the micro-mirrors and their corresponding reflection. The DMD is comprised of millions of micro-mirrors aligned in rows and columns. During the projection phase, the DMD mirrors would be either on or off, depending on the pattern being projected. The DMD mirrors that are turned on would absorb the UV light and project it downward, while mirrors that are turned off would reflect the UV light in the opposite direction [23]. Energy is

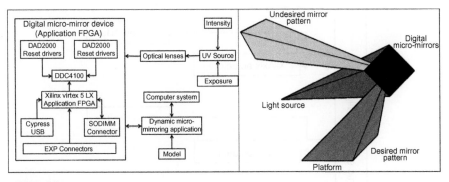

FIGURE 9.3

(Left) Structure of the digital micro-mirroring projection system. (Right) Illustration of light reflection on the digital mirrors.

lost during transmission of light from the UV source to the digital micro-mirror and from the micro-mirror to the substrate. The transmission wavelength of the system ranges from 370 nm to 410 nm, while the maximum energy output of the UV source is 15 MW at 100 percent intensity.

9.1.2 Photolithographic substrate alignment system

The photolithographic substrate alignment system consists of two actuators (manual): one that controls motion in the X direction (Cartesian coordinates) and one that controls motion in the Y direction (Cartesian coordinates). Together these two actuators span the X-Y plane of the alignment platform. Along with the X-Y actuators, there is a pair of fasteners that holds the substrate in place. Since this a microfabrication device, any small movement can be catastrophic. To ensure that the substrate's features are aligned with the projected mask (from the micro-mirror projection system), a digital microscopic device is directly below the alignment platform (centered) with a live feed to the computer system. This microscopic device features a fully adjustable magnification ranging from 20X to 200X. In addition to the magnification, the microscopic device has the capability to adjust its focal distance. The control structure of the photolithographic substrate alignment system is featured in Figure 9.4.

9.2 Multinozzle biologics deposition system

The multinozzle biologics deposition system is inspired by rapid prototyping technology and is built on the CAD/CAM platform, which is integrated with solid free-form automation to assemble biologics in three-dimensional space. This system consists of three motion arms for three-dimensional spatial control and a material deposition that houses up to 4 biological materials at once. The deposition system utilizes micro-valve nozzle systems that can deposit a wide range of

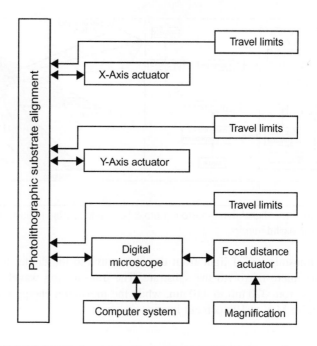

FIGURE 9.4

Structure of the photolithographic substrate alignment system.

solutions with a wide range of material and biological properties. This printer is fully integrated and computer controlled. The multinozzle biologics deposition system eliminates human error and provides its end users with precision control during fabrication procedures. This system executes micron-scale spatial control to generate cell-laden constructs. The multinozzle biologics deposition system is capable of depositing heterogeneous materials, cell types, and biological factors in a controlled and reproducible manner [24–26]. Cell printing is considered to be an effective biofabrication tool to assemble biologics. Figure 9.5 illustrates an overview of the multinozzle biologics deposition system configuration.

9.3 Methods and materials

The micro-fluidic chips are fabricated from two materials. Polydimethylsiloxane (PDMS) (Sigma-Aldrich, St. Louis, MO) is used as the basis for the chip, while SU-8 2100 (MicroChem Corp., Newton, MA) is used to fabricate the micro-architecture of the chip. We examine two biological explorations in this chapter. The first biological study is a preliminary investigation that characterized the fabricated chips in terms of the cell's ability to attach and proliferate in the micro-channels. The

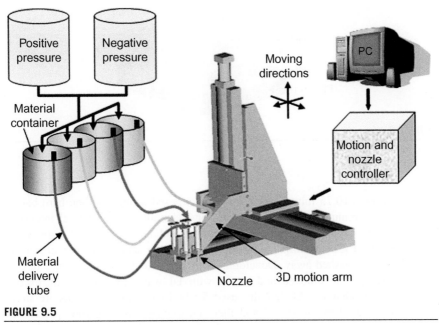

FIGURE 9.5

Overview of system configuration of 3D MDM system.

preliminary investigation used 7F2 cell line (mouse osteoblast), and the second biological investigation used MDA-MB-231 cell line (human breast cancer cells), both obtained from ATCC in Manassas, Virginia. The second cell line characterizes the model's potential to develop a cancerous model. (Unless listed otherwise, all cell culture supplements were obtained from ATCC.) All biological investigation data are expressed as the mean ± standard deviation for a sample size of 3 (n = 3).

9.3.1 Fabrication protocols and cell cultures

The enclosure of the chip is fabricated first. The enclosure has a platform (bottom) and a lid (top). The entire enclosure is fabricated from PDMS. The input and output ports are a nylon-based Luer-lock port (McMaster-Carr, Robbinsville, NJ). PDMS is mixed at 1:15 ratio, degassed, and cured in an aluminum mold at 130°C for 10 minutes. Cured PDMS is cooled and removed from the aluminum mold. This process is repeated for the lid, and the Luer-lock ports are placed in position prior to being cured on the hotplate. Figure 9.6 shows a model of the PDMS enclosure.

SU-8 is poured and leveled in the PDMS slot (bottom of the enclosure). It is then soft-baked at 65°C for 20 minutes, and then at 90°C for 220 minutes for stability. It is cooled for 30 minutes and then exposed at the recommended exposure time (this is based on the amount of energy required for crosslinking): DMM

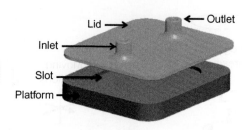

FIGURE 9.6

Model of the PDMS enclosure.

exposure time is 10.75 minutes, 265 mJ. The exposed sample is then hard-baked at 65°C for 15 minutes, and then at 90°C for 30 minutes for structural integrity. The sample is cooled for another 30 minutes and then developed with SU-8 Developer (MicroChem Corp., Newton, MA). During this stage, the unwanted material will be washed away. Development time is 8 to 15 minutes. Once developed, the sample is removed and rinsed with DI water to remove any excess materials in the channel. All samples used for biological investigations are sterilized first by dry heat of 150°C and then autoclaved. This rigorous process is used to ensure that the samples are completely sterile. Prior to cell deposition in the micro-channels, all samples are plasma treated with a Harrick Plasma Treater (Harrick Plasma, Ithaca, NY). Each sample was vacuumed to a pressure of 100 mTorr to 1 Torr, and the plasma was treated at high RF power settings for 120 seconds. Plasma treatment is used to created a seal between enclosures and to enhance cell attachment and proliferation with the SU-8 channels.

The DMM utilizes a systematic approach to fabricate models using a photosensitive polymer. The digital micro-mirroring microfabrication system projects an image of the desired structure onto the photosensitive polymer. Once the polymer is exposed, it manipulates the material's chemical properties and mimics the projected pattern [27]. Once exposed, the samples will undergo baking, developing, and sterilization processes (as previously detailed). Figure 9.7(a) shows the model of the microfluidic pattern, (b) the fabricated microfluidic pattern on the chip (bottom enclosure), and (c) a microscopic view of the channels in the chip.

7F2 and MDA-MB-231 cell lines have similar cell culture protocols. Both cell lines were seeded onto 75 cm^2 vented flasks and incubated. Six hours after the cells were seeded, the culture medium (cell depended) was changed to remove any dead cells in the flask; the culture medium was also changed every 2 to 3 days until the flasks were confluent. Confluent flasks were harvested and counted using a hemocytometer. The cells were then centrifuged again, in which the cell pallet was suspended to a cell density of 1×10^6 cells/mL. The cells were then loaded into the printer, where they were printed into the micro-channels (Figure 9.8). After the printing process, the chip would be sealed (sealing is possible to plasma treatment) with a PDMS lid, which has one input and one output. Culture media were changed

FIGURE 9.7

(a) Modeled microfluidic pattern. (b) Fabricated microfluidic pattern. (c) Microscopic view of the channel.

FIGURE 9.8

Close-up view of cells being printing into the micro-channels.

every 2 to 3 days and after every characterization point. 7F2 cells were incubated at 37°C with 95 percent air and 5 percent CO_2, while MDA-MB-231 cells were incubated at 37°C with 100 percent air.

9.4 Results

9.4.1 Preliminary investigation

The micro-channel array fabricated for this investigation is a single-layered open chip. An open chip study characterizes the feasibilities of the chip's potential for cell attachment and proliferation. The open chip study does not have a lid; cells are seeded into the micro-channels using the cell printer and then placed in the incubator for characterization. The two sample types for this preliminary investigation are nonplasma-treated micro-channels and plasma-treated micro-channels. Plasma treatment of the micro-channels enhances the surface properties. The chips fabricated for this investigation are the ones shown in Figure 9.7.

9.4.1.1 Cell interaction

It is important that the cells maintain interaction in the micro-channels; without the cell—cell interaction, cells cannot proliferate and differentiate into mature cells that are essential for functional tissues. A fluorometric indicator (Alamar Blue, Serotec) of cell metabolic activity was utilized to determine the cell proliferation in the channels [28—30]. Cell-laden samples (both treated and untreated) were removed from the culture plates, washed twice, placed in a new culture plate where 10 percent (v/v) Alamar Blue was added to the micro-channels, and incubated for 4 hours. The resulting solution was removed from the culture plate and placed in the microplate reader (GENios, TECAN, North Carolina) whose excitation and emission wavelengths were 535 nm and 590 nm, respectively. This study was conducted on days 0, 3, 5, 7, 10, and 14. The results of this study are shown in Figure 9.9, where the untreated samples showed no upregulation of cell proliferation. However, the plasma-treated samples showed a linear progression of proliferation up to day 7. After day 7, this progression subdued. The decline of proliferation of the plasma-treated samples on day 7 is believed to be due to limited space for cells to continue to grow in the micro-channels. Figure 9.10 is

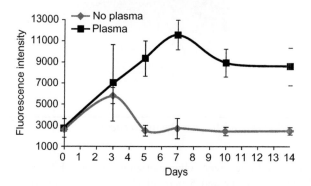

FIGURE 9.9

Cell proliferation on treated and untreated microfluidic constructs.

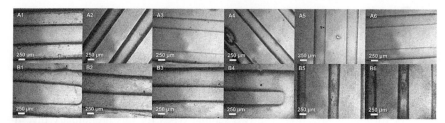

FIGURE 9.10

An optical snapshot of cells growing in the microfluidic chips on days 0, 3, 5, 7, 10, and 14 for untreated (A1—A6) and treated (B1—B6) samples.

an optical snapshot of cells growing in the microfluidic chips on days 0, 3, 5, 7, 10, and 14 for untreated (A1−A6) and treated (B1−B6) samples.

9.4.1.2 Cytotoxicity analysis

The MarkerGene^{TM} Live:Dead/Cytotoxicity Assay Kit (Marker Gene Technologies, Inc.; published by Marker Gene Technologies, Inc., Eugene, OR) was used to analyze the cytotoxicity of the plasma- and nonplasma-treated chips. Manufacturers' protocols were followed to create the working live:dead solution from the propidium iodide (PI) solution and the carboxyfluorescein di-acetate (CFDA) solution. The carboxyfluorescein dye is retained in live cells, producing a green fluorescence, while cells with damaged membranes allow the entrance of PI, which undergoes a fluorescence enhancement upon binding to nucleic acids, promoting a red fluorescence in dead cells. Qualitative and quantitative data were collected from the microfluidic cell-laden constructs on days 7 and 14 after cells were printed into the micro-channels. Figure 9.11 illustrates live (green) and dead (red) cells in the micro-channels of the construct. Images A1 and A2 in Figure 9.11 show the live and dead cells in the nonplasma-treated sample, while B1 and B2 show the live and dead cells in the plasma-treated micro-channels on days 7 and 14, respectively. On day 7, there were more living cells on the plasma-treated sample compared with day 14. Additionally, the plasma-treated samples show that its enhanced surface properties have the capability to support cell attachment and proliferation in the micro-channels.

9.4.2 Cancer model investigation

The preliminary investigation showed that chips fabricated with the DMM system have the potential to develop micro-tissue models. This investigation will move one step further and characterize the chip's ability to develop a cancer model. The chips used in this investigation are fully developed chips with a sealed

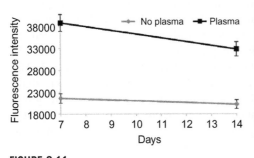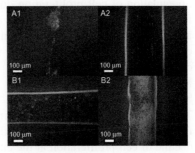

FIGURE 9.11

(Left) Quantitative results of live cells in the treated and untreated samples on days 7 and 14. (Right) Fluorescence images of cells in the channels of untreated (A1−A2) and treated (B1−B2) samples on days 7 and 14, respectively.

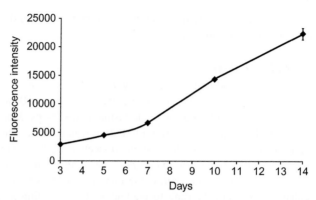

FIGURE 9.12

Cell proliferation of MDA-MB-231 cells in the micro-channels.

enclosure with one input and one output. All samples are plasma treated. The cells are printed into the micro-channels using the cell printer and are sealed immediately after printing is concluded. Culture medium is fed through the inlet of the chip according to the culture procedures discussed earlier in the chapter.

9.4.2.1 Cell interaction

A fluorometric indicator (Alamar Blue, Serotec) of cell metabolic activity was utilized to determine the cell proliferation in the channels. Cell-laden samples were sectioned, washed twice, and placed into a culture plate where 10% (v/v) Alamar Blue was added to micro-channels and incubated for 4 hours. The resulting solution was removed from the culture plate and placed into the microplate reader, whose excitation and emission wavelengths were 535 nm and 590 nm, respectively. This study was conducted for 14 days. The results of this study are shown in Figure 9.12, where the cells in the micro-channels showed a continuous upregulation of cell proliferation during the 14-days investigation. Data gathered from the preliminary study showed an upregulation and then a decline in proliferation after day 7. Data gathered in this study did not show the same trend. In comparison to the 7F2 cells used in the preliminary study, the MDA-MB-231 cells grow at a slower rate, hence the slow proliferation rate up to day 7. The MDA-MB-231 cells continued to proliferate for the remainder of the study because there was sufficient room for the cells in the micro-channels to grow. This was not the case for the preliminary investigation: the 7F2 cells grew at a fast rate, and at day 7 of the study they ran out of room to grow, hence the decline of cell proliferation at the end of the 14-day study.

9.4.2.2 Cytotoxicity analysis

The MarkerGene™ Live:Dead/Cytoxicity Assay Kit was used to provide qualitative data of cells in the micro-channels. Manufacturers' protocols were followed

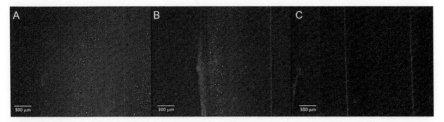

FIGURE 9.13

Live: Dead characterization of microfluidic chip of the MDA-MB-231 cell line on (A) day 7, (B) day 10, and (C) day 14.

to create the working live:dead solution from the propidium iodide (PI) solution and the carboxyfluorescein di-acetate (CFDA) solution. The carboxyfluorescein dye is retained in live cells, producing a green fluorescence, while cells with damaged membranes allow the entrance of PI, which undergoes a fluorescence enhancement upon binding to nucleic acids, promoting a red fluorescence in dead cells. Qualitative data were collected from the microfluidic cell-laden constructs on days 7, 10, and 14 after cells were printed in the channels. Figure 9.13 illustrates live (green) and dead (red) cells in the micro-channels of the construct. As seen in Figure 9.13, there are live (green glow) cells in the channels actively growing throughout the 14-day span.

9.4.2.3 Cell morphology

The micro-structural form of the micro-channels and cell morphology were evaluated using an FEI/Philips XL-30 Field Emission Environmental Scanning Electron Microscope (SEM). Images taken by SEM used a beam intensity of 2 KV and gaseous secondary electron detectors of 1.3 Torr. Before chips were characterized for architectural integrity and cell morphology, appropriate preparation was conducted by sectioning the lid of the enclosure. Sectioned samples were submerged in glutaraldehyde (GTA) (Sigma-Aldrich, St. Louis, MO) for 2 hours, followed by a dehydration process of submerging the GTA treated samples in 70%, 80%, 90%, 95%, and 100% ethanol for 10 minutes, respectively. Following the dehydration process, samples were placed under the culture hood for 1 hour to dry and then refrigerated overnight. Prior to SEM, samples were coated with platinum for enhanced visibility. As seen in Figure 9.14, channels are 300 µm wide and are uniform. Cells are attached in the channels and showed signs of proliferation. The morphology of the cells in the micro-channels is completely different from the morphology of cells in a two-dimensional culture. This may be due in part to the microfluidic environment [31−33].

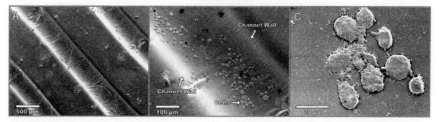

FIGURE 9.14

Scanning electron microscopy images showing (A) channel uniformity and orientation in the chip, (B) cell in the channel, and (C) close-up view of cells in the channel.

CONCLUSION

The DMM system has the capability to develop a biological microfluidic system. Chips fabricated with the DMM system have demonstrated their potential to promote cell attachment and proliferation. This preliminary investigation gives way for complex micro-architectural design for biological applications. The incorporation of a three-dimensional cell printer provided the added capabilities for precise spatial control of cells in the channels. Spatial orientation of cells will benefit the fabrication of complex future models; these models will be fabricated by a layer-by-layer technique. Additionally, the DMM system provides an economical fabrication technique to produce biological tissue arrays. The data presented show that the preliminary approach of developing a micro-platform to host cancerous cells is possible. The difference in cell morphology (two-dimensional vs. microfluidic) demonstrates that the micro-environment has a great influence on the cancer cells. A three-dimensional cancer model will closely mimic the in vivo conditions, thus providing researchers with an additional method of characterization pharmaceuticals.

References

[1] Starly B. Biomimetic design and fabrication of tissue engineered scaffolds using computer aided tissue engineering. Philadelphia, PA, USA: Drexel University; 2006 [Ph.D. Dissertation, Mechanical Engineering and Mechanics].

[2] Shor L. Novel fabrication development for the application of polycaprolactone and composite polycaprolactone/hydroxyapotote scaffolds for bone tissue engineering. Philadelphia, PA, USA: Drexel University; 2008 [Ph.D. Dissertation, Mechanical Engineering and Mechanics].

[3] Huang GY, Zhou LH, Zhang QC, Chen YM, Sun W, et al. Microfluidic hydrogels for tissue engineering. Biofabrication 2011;3:012001.

[4] Jo BH, Van Lerberghe LM, Motsegood KM, Beebe DJ. Three-dimensional micro-channel fabrication in polydimethylsiloxane (PDMS) elastomer. J Microelectromech Syst 2000;9:76−81.

[5] Ho CM, Tai YC. Micro-electro-mechanical-systems (MEMS) and fluid flows. Annu Rev Fluid Mech 1998;30:579−612.

[6] Spearing SM. Materials issues in microelectromechanical systems (MEMS). Acta Mater 2000;48:179−96.

[7] The Organ and Transplantation Network. Available: <http://www.ustransplant.org>; 2004.

[8] Parnes LS, Sun AH, Freeman DJ. Corticosteroid pharmacokinetics in the inner ear fluids: an animal study followed by clinical application. Laryngoscope 1999;109:1−17.

[9] Elliott NT, Yuan F. A review of three-dimensional in vitro tissue models for drug discovery and transport studies. J Pharm Sci 2011;100:59−74.

[10] Zhang X, Wang W, Yu W, Xie Y, Zhang X, et al. Development of an in vitro multi-cellular tumor spheroid model using microencapsulation and its application in anti-cancer drug screening and testing. Biotechnol Prog 2005;21:1289−96.

[11] Hassan SB, de la Torre M, Nygren P, Karlsson MO, Larsson R, et al. A hollow fiber model for in vitro studies of cytotoxic compounds: activity of the cyanoguanidine CHS 828. Anticancer Drugs 2001;12:33−42.

[12] Cowan DS, Hicks KO, Wilson WR. Multicellular membranes as an in vitro model for extravascular diffusion in tumours. Br J Cancer Suppl 1996;27:S28−31.

[13] Casciari JJ, Hollingshead MG, Alley MC, Mayo JG, Malspeis L, et al. Growth and chemotherapeutic response of cells in a hollow-fiber in-vitro solid tumor-model. J Natl Cancer Inst 21 1994;86:1846−52.

[14] Friedrich MJ. Studying cancer in 3 dimensions—3-D models foster new insights into tumorigenesis. JAMA 2003;290:1977−9.

[15] Xiang D, Arnold MA. Solid-state digital micro-mirror array spectrometer for Hadamard transform measurements of glucose and lactate in aqueous solutions. Appl Spectrosc 2011;65:1170−80.

[16] Adeyemi AA, Barakat N, Darcie TE. Applications of digital micro-mirror devices to digital optical microscope dynamic range enhancement. Opt Express 2009;17:1831−43.

[17] Shin W, Yu BA, Lee YL, Yu TJ, Eom TJ, et al. Tunable Q-switched erbium-doped fiber laser based on digital micro-mirror array. Opt Express 2006;14:5356−64.

[18] Lu Y, Mapili G, Suhali G, Chen S, Roy K. A digital micro-mirror device-based system for the microfabrication of complex, spatially patterned tissue engineering scaffolds. J Biomed Mater Res A 2006;77:396−405.

[19] Gauvin R, Chen YC, Lee JW, Soman P, Zorlutuna P, et al. Microfabrication of com-plex porous tissue engineering scaffolds using 3D projection stereolithography. Biomaterials 2012;33:3824−34.

[20] Catros S, Guillemot F, Nandakumar A, Ziane S, Moroni L, et al. Layer-by-layer tis-sue microfabrication supports cell proliferation in vitro and in vivo. Tissue Eng Part C Methods 2012;18:62−70.

[21] Andersson H, van den Berg A. Microfabrication and microfluidics for tissue engi-neering: state of the art and future opportunities. Lab Chip 2004;4:98−103.

[22] Starly B, Sun W. Internal scaffold architecture designs using Lindenmayer systems. J Comput Aided Des Appl 2007;4:395−403.

[23] Nederman T, Acker H, Carlsson J. Penetration of substances into tumor tissue: a methodological study with microelectrodes and cellular spheroids. In Vitro 1983;19:479–88.

[24] Chang R, Sun W. Biofabrication of three-dimensional liver cell-embedded tissue constructs for in vitro drug metabolism models. LAP Lambert Academic Publishing; 2009.

[25] Sun W, Darling A, Starly B, Nam J. Computer-aided tissue engineering: overview, scope and challenges. Biotechnol Appl Biochem 2004;39:29–47.

[26] Sun W, Lal P. Recent development on computer aided tissue engineering–a review. Comput Methods Programs Biomed 2002;67:85–103.

[27] Guijt RM, Breadmore MC. Maskless photolithography using UV LEDs. Lab Chip 2008;8:1402–4.

[28] Hamid Q, Snyder J, Wang C, Timmer M, Hammer J, et al. Fabrication of three-dimensional scaffolds using precision extrusion deposition with an assisted cooling device. Biofabrication 2011;3:034109.

[29] Shor L, Guceri S, Chang R, Gordon J, Kang Q, et al. Precision extruding deposition (PED) fabrication of polycaprolactone (PCL) scaffolds for bone tissue engineering. Biofabrication 2009;1.

[30] Yan KC, Nair K, Sun W. Three dimensional multi-scale modelling and analysis of cell damage in cell-encapsulated alginate constructs. J Biomech 2010;43:1031–8.

[31] Hung PJ, Lee PJ, Sabounchi P, Lin R, Lee LP. Continuous perfusion microfluidic cell culture array for high-throughput cell-based assays. Biotechnol Bioeng 2005;89:1–8.

[32] Tourovskaia A, Figueroa-Masot X, Folch A. Differentiation-on-a-chip: a microfluidic platform for long-term cell culture studies. Lab Chip 2005;5:14–9.

[33] Kim MS, Yeon JH, Park JK. A microfluidic platform for 3-dimensional cell culture and cell-based assays. Biomed Microdevices 2007;9:25–34.

Breast Reconstruction Using Biofabrication-Based Tissue Engineering Strategies

10

Mohit P. Chhaya[1], Ferry P.W. Melchels[1,2], Paul S. Wiggenhauser[3], Jan T. Schantz[3], and Dietmar W. Hutmacher[1,4]

[1]Institute of Health and Biomedical Innovation, Queensland University of Technology, Brisbane, Australia; [2]Department of Orthopaedics, University Medical Center, Utrecht, the Netherlands; [3]Department of Plastic and Hand Surgery, Klinikum Rechts der Isar, Technical University Munich, Germany; [4]George W. Woodruff School of Mechanical Engineering, Georgia Institute of Technology, Atlanta, Georgia, USA

CONTENTS

Biofabrication.

INTRODUCTION

Breast cancer is a major cause of death in women. In 2000, 375,000 women died from breast cancer worldwide [1]. The most common surgical procedures to remove the tumor are lumpectomy, which is removal of the tumor and surrounding breast tissue, and total mastectomy, which is the total removal of the breast. Such procedures have a negative psychological effect on the well-being of the patient [2]. Earlier studies done by Renneker et al. show that mastectomy is directly related to a psychological syndrome "marked by anxiety, insomnia, depressive attitudes, occasional ideas of suicide, and feelings of shame and worthlessness" [2].

Owing to the large number of surgeries, breast reconstruction following such procedures is becoming increasingly common. In 2011, about 307,180 breast reconstruction procedures were performed in the United States alone [3].

10.1 Anatomy of the breast

To appreciate the engineering challenges associated with breast reconstruction, it is necessary to understand the anatomy involved. The breasts are located on the anterior and lateral parts of the chest, and their primary role is to provide milk for the nourishment of the infant [4]. The two parts of the breast are the internal area and the external area. The external parts include the nipple, areola and tubercles [4]. The internal part, which is also the principal secretory organ, is assembled by 15 to 25 lobes of compound milk-producing glands embedded in fibrous and adipose tissue [5]. Each of these lobes contains an excretory duct that drains into lactiferous sinus.

The adipose tissue interspersed among these glands is perhaps the most important factor determining the shape and volume of the breast [4,6,7]. It also plays a major role in determining the softness and contour of the breast [6].

The breast is attached to the underlying structures with fibrous bands known as suspensory (or Cooper's) ligaments [6]. These ligaments also play a role in determining the shape and contour of the breast [7]. Figure 10.1 shows the anatomy of the breast.

10.2 Current methods of breast reconstruction

Currently, the three most common surgical approaches for reconstructive surgery after lumpectomy or mastectomy are reconstruction with autologous tissue using silicone implants, free/pedicled flaps, or lipofilling.

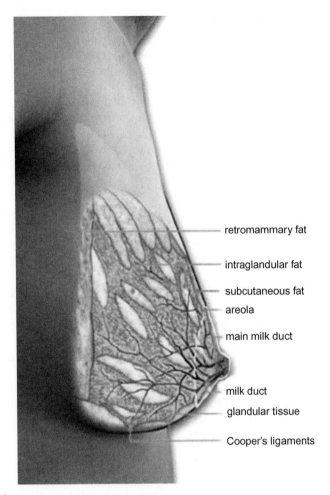

retromammary fat

intraglandular fat

subcutaneous fat

areola

main milk duct

milk duct

glandular tissue

Cooper's ligaments

FIGURE 10.1

The gross anatomy of the lactating breast based on ultrasound images of the milk duct system and distribution of different tissues in the breast.

Source: From [8].

10.2.1 **Prosthetic implant-based reconstruction**

This relatively easy surgical approach relies on the implantation of prosthetic devices, such as fixed-volume breast implants and tissue expanders. The advantage of using such devices is that they can be manufactured in a broad range of sizes, contours, profiles, and textures [6]. The two major types of prosthetic implants are fixed volume implants and tissue expanders.

10.2.1.1 Fixed volume implant

The fixed-volume breast implant is a single lumen implant made of silicone elastomer that is filled with a fixed volume of saline solution during the surgery [9]. Adjustments cannot be made to the saline volume after the operation. The saline solution can also be replaced with silicone. However, controversy still exists with regard to silicone's association with numerous health problems [10]. Moreover, studies by Flassbeck et al. have found that siloxanes and platinum could leak out of such implants, and their levels were elevated in the fatty tissues of women with leaking implants [11]. More studies, however, are needed to establish a link between elevated siloxane levels and health problems in women with implants.

10.2.1.2 Tissue expanders

A tissue expander is like an inflatable breast implant. It is inserted into the breast in its collapsed form during the surgical procedure and is gradually inflated by the injection of saline over the course of weeks to months [9]. When the expansion is complete, it is either replaced with a permanent saline- or silicone-based implant or is left inside.

Perhaps the biggest disadvantage with using prosthetic implants is that of capsular contracture. Multiple studies have demonstrated that inserting such an implant leads to a foreign body reaction, resulting into the formation of a capsule of fibrous tissue around the implant [12−15]. This ultimately leads to a spherical appearance of the breast and restricted shoulder or arm movement [16]. The frequency of occurrence of capsular contracture is from 2 percent to 70 percent, depending on the study and the patient cohort involved [17−20]. There is an average reported capsular contracture risk of approximately 10 percent. This risk of capsular contracture also rises significantly when the breasts are irradiated following the implantation [21−24]. Figure 10.2 shows examples of capsular contracture around breast implants.

In addition, both types of implants are also subject to possible rupture, displacement, deformation, chronic seroma, hematoma, and loss of nipple sensation [9,15,26]. Therefore, breast reconstruction using implants is not a completely sustainable solution.

10.2.2 Cellular breast reconstruction

Cellular breast reconstruction does not rely on prosthetic implants. Currently, two methods of cellular reconstruction seem promising: autologous fat transplant and de novo adipose tissue engineering [27]. While only autologous fat transplant is used today, numerous in vivo proof-of-concept studies have been conducted to determine if the method is preferable.

10.2.2.1 Autologous fat transplant

Autologous fat transplant involves using liposuction to transfer fat from a donor site in the patient's body to the breast region in the form of hundreds of tiny

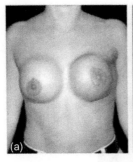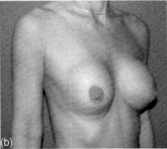

FIGURE 10.2

Examples of capsular contracture around breast implants. (A) Patient possesses capsular contracture of both breasts, resulting in breast hardness, an unnatural spherical profile, and breast asymmetry. (B) Similarly, this patient possesses capsular contracture of her right breast.

Source: Images taken from Ma et al. [25]. Reproduced with permission.

droplets called a lipoaspirate [28]. The full breast can then be reconstructed with repeated sessions of fat transfer.

However, autologous fat transplantations yield poor results, with 40 to 60 percent reduction in graft volume owing to tissue resorption and necrosis [29−31]. Insufficient vascularization is also thought to be one of the causes leading to such a reduction in adipose volume.

10.2.2.2 Autologous fat transfer with the BRAVA system

In this emerging method of breast reconstruction, a proprietary system called BRAVA places a gentle amount of pressure on the breast that, when sustained, causes the cells to respond by replicating and causes the blood vessels to expand [32]. In a successive step, fat is harvested from several areas of the patient's body via liposuction and is injected into the breasts. Because of the increased volume and vascularization created by the BRAVA system, a significant number of new vessels form around the new fat deposits, resulting in better graft survival rates [32].

Although this system is still awaiting FDA approval, if used, it should help the survival of fat grafts by providing increased vascularization and volume. Figure 10.3 shows the application of the BRAVA system.

10.2.2.3 Free tissue transfer flaps

Using transfer flaps is similar to autologous fat transfer. The difference between the two techniques is that in the case of free tissue transfer flaps, the tissue is transferred along with its blood vessels [33]. The blood vessels in the flap are then connected to the vessels at the recipient site.

Several sites can be used for flap tissue, but the transverse rectus abdominis musculocutaneous (TRAM) flap described by Hartrampf et al. [34] and the deep inferior epigastric perforator (DIEP) flap described by Allen et al. [35] are

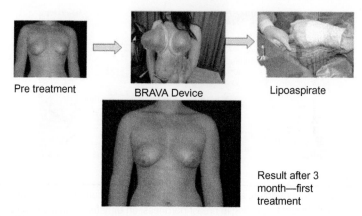

Pre treatment   BRAVA Device   Lipoaspirate

Result after 3
month—first
treatment

FIGURE 10.3

The BRAVA system and lipofilling for augmentation of a tuberous breast deformity.

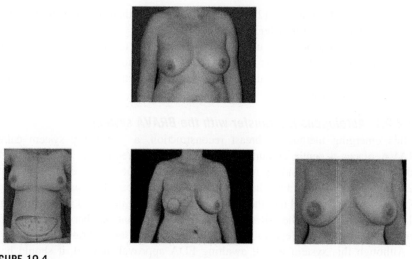

FIGURE 10.4

Breast reconstruction using the DIEP flap.

currently the gold standards in breast reconstructive surgery. Figure 10.4 shows a breast reconstruction procedure done using the DIEP flap.

The main complication arising with free tissue transfer flaps is that sometimes a clot forms in the vein that drains blood from the flap or the artery that supplies blood to the flap [36,37]. Both cases may lead to necrosis of the flap tissue. Other inherent risks include total/partial flap loss and abdominal bulge or hernia [38]. The incidence of complications after TRAM flap reconstruction in the breast region ranges from 8 to 13 percent, while the incidence of complications in the breast region ranges from 1 to 82 percent [39−46]. The DIEP flap also suffers breast-related morbidity issues, including a fat necrosis ranging from 6 to 62.5 percent [35,37,47,48]. Nahabedian et al. [38] provides an excellent review on this topic.

10.2.2.4 *De novo adipose tissue engineering*

Recent progress in the field of tissue engineering has been made by seeding human adipose tissue−derived precursor cells (hAPCs)—fibroblast-like precursor cells that differentiate into mature adipocytes—onto a biodegradable scaffold, where they promote the formation of adipose tissue [49,50]. The main advantage of de novo tissue engineering is that the scaffolds degrade in vivo, thus allowing for remodeling of the tissue without the long-term presence of foreign material. Furthermore, this method does not suffer from the shrinkage of graft volume [27] and does not suffer as severe complications as prosthetic implants. hAPCs, which are also known as *preadipocytes*, can be easily cultured using standard techniques, and indeed many groups have demonstrated successful isolation and culturing of human, rat, and swine preadipocytes [51−53].

De novo adipose tissue engineering is generally undertaken with the help of tissue engineered constructs that consist of three major components: cells, a biodegradable scaffold, and a microenvironment suitable for cellular growth and differentiation [54]. When implanted into the test subject, these constructs initiate and direct the formation of de novo tissue. Over time, the scaffold degrades and the newly formed tissue takes its place.

10.3 Biofabrication-based tissue engineering strategy

10.3.1 Engineering challenges

The overall goal of breast reconstruction is to restore the patient's breast mass with adipose tissue and maintain tactile sensation. Every patient's breast will have a unique size and shape, so the tissue construct used for adipose tissue engineering needs to be highly customized. Also, research has shown that the breast's architecture with regard to adipose tissue volume and skin elasticity and thickness changes with time [7]. The tissue construct should adapt with such changes. Moreover, the scaffold used in the tissue construct must be biodegradable and must not need surgical removal. It should also not invoke strong inflammatory response or long-term fibrous encapsulation.

To develop a customized breast scaffold, an integrated approach needs to be used that links a laser scanner with computer-aided design and computer-aided manufacturing (CAD/CAM). These features are explored in the following.

10.3.1.1 Imaging

The first step toward creating a customized breast scaffold is to obtain data related to the shape and size of the patient's breast.

10.3.1.1.1 Laser digitizer

A laser digitizer can be used to scan the breast region of the healthy breast and capture the point data based on the principle of laser triangulation [55]. Precise scanning of the breast surface has already been performed extensively by Kovacs et al. [56,57]. This technique can also be used to obtain patient-specific 3D data on breast shape and size with accuracy comparable to that of MRI scanning [58] that can then be used to fabricate patient-specific scaffolds.

10.3.1.1.2 Bioluminescence imaging

Bioluminescence refers to the enzymatic generation of visible light by living organisms by using a bioluminescent reporter gene such as luciferase. Using bioluminescence as an imaging technique for in vivo models is described in detail by Edinger et al. [59]. First of all, the bioluminescent reporter gene needs to be transferred to the protein of interest using standard gene transfer methods [60–62]. Cells with stable expression of this gene can then be injected into the animal of interest, and the light emitted from the cells can be monitored externally by placing the animals inside a light-tight chamber with a charge-coupled device camera. The data captured by the camera are then transferred to a computer equipped with image acquisition and analysis software [63].

10.3.1.1.3 Single photon emission computed tomography

Single photon emission computed tomography (SPECT) techniques require the injection of a gamma-emitting radioisotope such as ^{125}I into the body of the subject [64]. A scintillation camera then captures multiple 2D images of the 3D distribution of the radioisotope from multiple angles. These images are transferred to a computer that can be used to apply a tomographic reconstruction algorithm to form a 3D model that can be used to fabricate tissue engineering scaffolds [65–67].

10.3.1.1.4 Magnetic resonance imaging

A magnetic resonance imaging (MRI) scanner uses a powerful magnet to align the magnetization of atomic nuclei in the body. It then uses radio frequencies to alter the alignment of the magnetization, causing the nuclei to produce a rotating magnetic field that can be detected by the scanner [68]. The information obtained from the rotating magnetic field can then be used to generate 2D or 3D images of the scanned area. MRI has already been used for 3D scanning of the knee joint [69] and thus holds promise in the area of breast imaging as well.

10.3.1.1.5 Computed tomography

In contrast to MRI scanning, a computed tomography (CT) scanner uses computer-processed x-rays to produce "slices" of specific areas of the body that can then be fed into a tomographic reconstruction algorithm similar to SPECT to generate 3D images [66]. However, since CT scanning uses x-rays, there is a risk that it may damage the cellular DNA in the body part that is irradiated, which may lead to a lifetime attributable risk of cancer [70,71].

10.3.1.2 Triangulated surface model

After obtaining the 3D image, several modeling steps need to be undertaken to obtain a computer model of the breast. As described by Kovacs et al. [56], the borders of the breast region need to be defined at 1 cm below the clavicle to 1 cm below the lower breast fold. A surface interpolation algorithm then needs to be applied between these two borders to generate a virtual chest wall [72].

The point data obtained from the laser scan is then translated into a CAD model and saved as a Standard Tessellation Language (STL) file. The basic algorithm for performing this task is described in detail by Chua et al. [73]. In short, the triangulated surface is generated by connecting all the scanned points to form small triangles. The model obtained using the preceding method is then excised at the breast boundary defined in the previous paragraph to obtain a final CAD model that can be used for scaffold fabrication. Such a model is shown in Figure 10.5.

It is important to note here that the model obtained is for a healthy breast. Therefore, in the CAD software, the diseased breast must be virtually removed and replaced by the modeled chest wall for the healthy region to ensure a good fit [72].

10.3.1.3 Scaffold design and porosity

Since a CAD model is essentially a solid model, it needs to be modified to generate porosity within the scaffold. Melchels et al. [72] describe a method to generate porosity within a solid model using finite element analysis to generate tetrahedral

 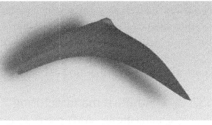

FIGURE 10.5

CAD model of a healthy breast obtained using a laser scanner.

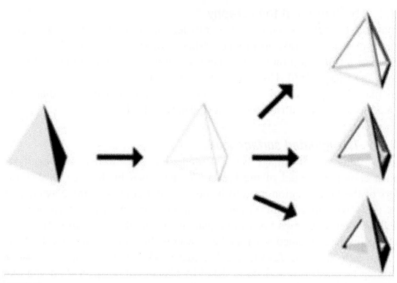

FIGURE 10.6

Generation of porosity on a solid model. A solid model of a tetrahedron is first created using finite element analysis. A custom algorithm was then designed to create struts of defined thickness along the edges of the vertices.

Source: Image taken from Melchels et al. [72]. Reproduced with permission.

volumetric meshes as intermediates as a first step. In the second step, they designed an algorithm to create struts of defined thickness around each edge length of all tetrahedrons. The tetrahedrons were then joined at the intersections. These steps are then repeated for all individual elements, resulting in a porous interconnected final structure. Figures 10.6 and 10.7 illustrate this process.

10.3.1.4 *Scaffold manufacturing*

Several technologies have been developed to fabricate tissue engineering scaffolds. These include solvent casting, particulate leaching, emulsion freeze-drying, thermally induced phase separation, and so on. However, all of these techniques suffer from issues such as low pore interconnectivity [75], use of toxic solvents [75], residual porogen particles in the polymer matrix [76], and inconsistency in the pore sizes and distribution [77].

10.3.1.4.1 Additive manufacturing

Additive manufacturing (AM) refers to a set of a techniques used to fabricate a physical model using three-dimensional computer-aided design (CAD) data. As opposed to the subtractive manner of manufacturing that conventional machining is based on, additive manufacturing operates by fusing material in a layer-by-layer fashion. Such a building process ensures minimal waste of raw material

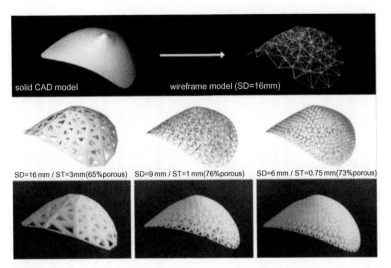

FIGURE 10.7

Generation of porous structures from a solid breast model. Scaffolds are based on meshes with three different seeding densities (SD—16, 9, and 6 mm distance—each for two different values of strut thickness (ST). The models were prototyped using fused deposition modeling.

Source: Figure taken from Melchels et al. [74]. Reproduced with permission.

such as cells, biomaterials, and growth factors. In contrast to the conventional scaffold manufacturing techniques, AM has the potential to design bioabsorbable scaffolds with a fully interconnected pore network that mimic the microstructure of living tissue [78,79]. AM also generally does not require the use of toxic solvents and offers great ease and flexibility in material handling [80]. It can also be used to create scaffolds with controlled porosity and a fully interconnected pore network, which is crucial for tissue ingrowth [78,80,81].

Several commonly used AM techniques are available commercially, such as stereolithography, inkjet printing, selective laser sintering (SLS), and fused depositional modeling. Such techniques manipulate materials in one of several possible ways: thermal, chemical, mechanical, and/or optical. In thermal processes like SLS, a heat source such as a high-powered laser beam fuses a layer of powdered biomaterial into a mass that has a predefined three-dimensional shape obtained from a CAD file. In chemical-based processes, the final shape of the prototype is fixed by a chemical reaction (e.g., polymerization). Mechanical techniques such as fused depositional modeling, on the other hand, rely on physical deposition of material in a layer-by-layer fashion, while optical processes such as stereolithography are based on the photopolymerization of a liquid polymer material with the help of a UV laser. Table 10.1 describes four common commercially available AM techniques that can be used to fabricate tissue engineering scaffolds.

Table 10.1 Commercially Available AM Techniques

Stereolithography A computer-controlled laser beam illuminates a pattern on the surface of a resin reservoir, which causes the resin to solidify to a defined depth. Once this process is completed, the stage is moved away from the surface and the built layer is recoated with liquid resin. A pattern is then cured in this second layer.

This photopolymerization process is repeated to generate subsequent layers. After the fabrication is completed, the excess resin is drained from the bath and the model is cured in a UV oven to improve its mechanical properties [82].

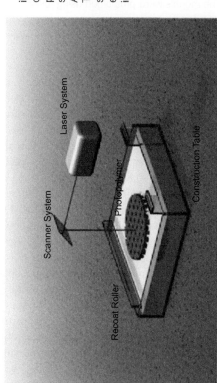

Inkjet Printing This process deposits a stream of microparticles of a binder material (or "ink") onto the surface of a powder bed (or "paper"). The printer then joins the particles together based on the data obtained from a CAD file. Once a layer has been deposited, the powder bed is lowered by a precise distance and a new layer is deposited and fused to the previous layer. This process is repeated until the 3D object is completely formed.

Using this technique, viable cells can be delivered to precise target positions on scaffold biomaterials [83]. Moreover, by delivering different cell types to different positions, tissue structures of the original structures can be precisely mimicked [84].

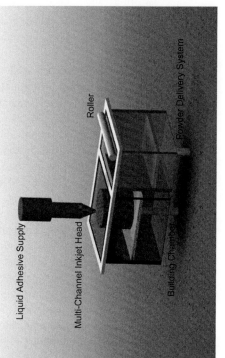

Selective Laser Sintering This technique uses a CO_2 laser beam that fuses a layer of powdered biomaterial into a mass that has a predefined three-dimensional shape obtained from a CAD file [85]. During the fabrication process, the laser is scanned over the powder surface following the geometry obtained from the CAD data. Focusing of the laser beam causes the temperature of the powder to rise, and sintering occurs at just beyond the glass transition temperature, causing the particles to fuse together into a solid mass [79]. After each layer, the powder bed is lowered by a predefined thickness and a new layer of material is applied on top, thus fabricating the next layer. The model is then created in a layer-by-layer fashion.

Fused Deposition Modeling The FDM method forms 3D objects from computer-aided design models derived from computer tomography scans or 3D digitizer systems. FDM machines use a temperature-controlled extruder to force out a thermoplastic filament and deposit the semimolten polymer onto a stage in a layer-by-layer fashion [78,81]. At the end of each layer, the extruder moves upward by a predefined amount, and the next layer is deposited.

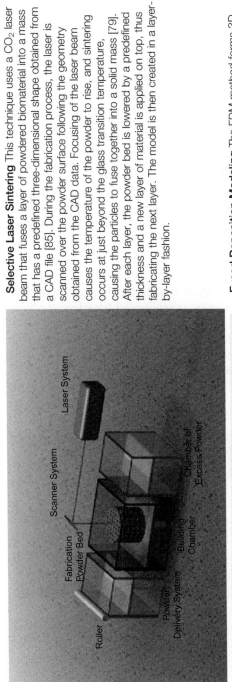

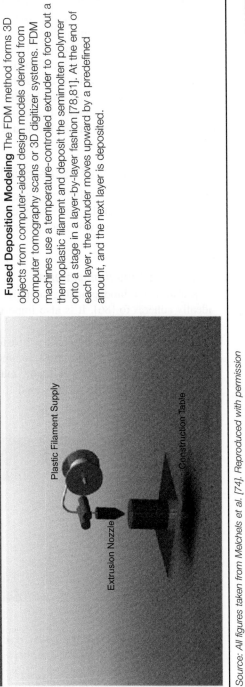

More recently, AM has also been used in conjunction with the conventional free tissue transfer flap surgery. Melchels et al. [72] used a solid CAD model of the breast shown in Figure 10.8 to fabricate an acrylic mold using a 3D printer. The mold was then used during a breast reconstruction procedure to orientate and contour the flap after translocation to the recipient site. The mold was also very helpful in shaping the breast, resulting in a more precise shape and a higher degree of geometry. It can thus be shown that AM can already be used to generate better outcomes in breast reconstruction surgeries employing the current gold standard of autologous breast reconstruction.

10.3.2 Formation of tissue constructs

The scaffolds formed using the preceding process can then be seeded with hAPC cells as described in the previous section and can be implanted into test subjects along with a suitable vascularization strategy. The entire strategy is summarized in Figure 10.9.

10.3.2.1 Scaffold biomaterial

A scaffold is a biodegradable support structure that the newly engineered tissue ideally adheres to [86]. It is also important for providing the boundary conditions for the overall tissue shape [54].

Many materials have been used to form such scaffolds, but porous biodegradable polymer foams are the most common [31,87]. However, Patrick points out in his review that such polymer foams may not be the best choice for use in breast tissue engineering because they tend to be very rigid and may be uncomfortable for the patient [54].

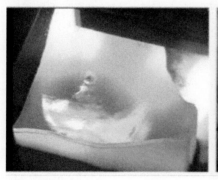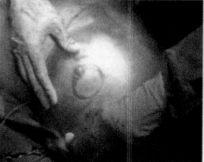

FIGURE 10.8

Left: Transparent acrylic mold used for mold-assisted breast reconstruction surgery. Right: The intraoperative use of the mold to shape the breast in flap transplantation-reconstruction.

Source: Figure taken from Melchels et al. [72]. Reproduced with permission.

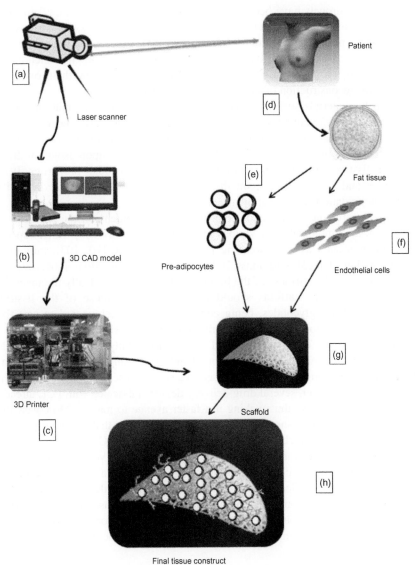

FIGURE 10.9

Tissue engineering strategy for breast reconstruction. First, an image is taken of the chest region using a 3D laser digitizer (a). This image is then processed into a CAD file (b), which is then converted into gcode, and the scaffold is finally fabricated using the BioExtruder (c). On the tissue culture side, fat tissue is harvested from the patient (d). Preadipocytes and endothelial cells are then separated from it (e, f) and finally cocultured onto the fabricated scaffold (g). This tissue engineering construct (h) is then implanted into an in vivo animal model.

Injectable hydrogels have been investigated by several groups [88,89] for adipose tissue engineering; however, such hydrogels, by themselves, do not possess the mechanical properties required to hold the shape of the breast mound [6,54]. Moreover, their low moduli do not provide enough support for anchorage-dependent cells [90]. Because of these reasons, hydrogels on their own are unsuitable for reconstruction of large volumes of tissue.

Natural biomaterials such as collagen [91,92] and synthetic biomaterials such as polylactide, polyglycolide, and their copolymers [93−96] have received much attention as potential candidates for tissue engineering scaffolds. However, collagen suffers from the problem of batch-to-batch variation upon isolation from biological tissues and does not provide tunable mechanical properties [97]. Polylactides and polyglycolides, on the other hand, are stiff biomaterials with very high degradation rates, as shown in Table 10.2.

The differentiation of precursor cells is highly dependent on the stiffness of the scaffold, as demonstrated by various research groups [98−100]. Other aspects of cell function such as cell spreading [101], proliferation [102,103], migration [104], formation of focal adhesions [105], and protein expressions [106,107] are also influenced by the stiffness of the biomaterials. Therefore, ideally the mechanical properties of the scaffolds should closely resemble those of the tissues. Polylactides and polyglycolides, being considerably stiffer compared to soft tissues, cannot be used for optimum regeneration of breast tissue.

A recent study by Wiggenhauser et al. [114] successfully used scaffolds made from PCL for adipose engineering. A major advantage of using PCL is that it is relatively inexpensive to produce and can be used in a variety of scaffold fabrication technologies [115]. Moreover, several drug-delivery devices fabricated with PCL already have FDA approval [115], thus making it a faster avenue to market. Moreover, the Young's modulus of PCL, shown in Table 10.3, is similar to that of the suspensory ligaments in the breast, which have a moduli range of 0.08−0.4 GPa [113].

However, the degradation time of PCL is approximately 2 years [116], whereas 6 months would be a sufficient time for the development of adipose tissue [117]. Therefore, chemical modifications need to be done to PCL to decrease its degradation time and stiffness before it can be optimally used for breast tissue engineering [118]. Also, PCL is hydrophobic, which makes it unsuitable for cell growth [119], so several attempts have been made to modify its surface to increase the cytocompability [119−121].

Table 10.2 Mechanical Properties of Commonly Used Biomaterials

Biomaterial	Bulk Elastic Modulus (kPa)
Poly-D,L-lactide (PDLLA)	$1.5-1.9 \times 10^6$ [108,109]
Poly-L-lactide (PLLA)	$1.4-2.0 \times 10^6$ [109−111]
Poly-caprolactone	0.31×10^6 [112]

10.3.2.1.1 Biodegradable elastomers

Elastomers are materials that exhibit high extensibility and complete shape recoverability [122]. Biodegradable elastomers are gradually gaining interest in the field of regenerative therapies for the purposes of fabricating flexible tissue engineering scaffolds and vessels for drug delivery systems [123–125]. Table 10.4, which is adapted from Shi et al. [125], summarizes the different types of biodegradable elastomers that have been investigated for tissue engineering purposes.

As can be seen from Table 10.4, elastomeric biopolymers have considerably lower bulk tensile modulus compared to stiff polymers such as PCL and polylactides. In this respect, it can be argued that such elastomers, perhaps in combination with hydrogels, are the most suited biomaterials for soft tissue engineering.

10.3.2.1.2 Hydrogels

Many gels have been investigated for tissue engineering applications. Native extracellular matrix (ECM) molecules such as collagen have already been investigated to create cell-laden hydrogels; however, their applications in terms of large-volume, long-term tissue engineering are limited due to poor mechanical properties [154]. Synthetic hydrogels, on the other hand, such as polyethylene glycol (PEG) [155,156] or hyaluronic acid (HA) [156,157], have stronger mechanical properties and exceptional encapsulated cell viability; however, cells find it difficult to bind to or degrade such gels [154].

One emerging hydrogel that can potentially overcome the limitations of natural and synthetic hydrogels is gelatin methacrylate (GelMA). GelMA is a photopolymerizable hydrogel comprised of gelatin, or denatured collagen, that is naturally present in ECM components [158]. The gelatin used to formulate GelMA also retains natural cell binding motifs, such as RGD [159] and MMP-sensitive

---

Table 10.3 Mechanical Properties of Tissue Components of the Breast

Tissue Type	Elastic Modulus (kPa)
Ribs	$2-14 \times 10^6$
Pectoralis major and minor muscles	*In the longitudinal direction:* For dynamic loading: ~ 30 *In the transverse direction:* For dynamic loading: 1.5–6
Pectoralis fascia	100–2000
Suspensory ligaments	80,000–400,000
Glandular tissue	7.5–66
Adipose	0.5–25
Skin	200–3000

Source: Adapted from Gefen et al. [113]. Reproduced with permission.

Table 10.4 Mechanical Properties of Elastomeric Biomaterials

Bioelastomer	Glass Transition Temperature (°C)	Tensile Strength (MPa)	Elongation at Break (%)	Approximate Degradation Time (Months)	Tested for Biofabrication	Reference
Polyurethanes	−116 to −41	4–60	100–950	Variable	Yes [126]	[127,128]
Polyphosphazenes	−105 to 91	Wide range	Wide range	Variable	Yes [129]	[130,131]
Poly(glycerol sebacate)	−7 to 46	>0.5	>267	1	Yes [132]	[133,134]
Poly(1,8-octanediol-co-citric acid)	−5 to 10	Up to 6.7	265 ± 10	Variable	Yes [135]	[136]
Poly(1,10-decanediol-co-D,L-lactic acid)	−5 to 10	Up to 3.14 ± 0.5	322 ± 20	Variable	-	[137]
Poly(diol citrates)	−5 to 10	Up to 11.2	Up to 502	Variable	-	[137]
Poly(ethylene glycol)/poly(butylene terephthalate)	—	8 to 23	500 to 1300	Variable	Yes [138–140]	[141,142]
Poly(glycolide-co-caprolactone)	—	<1	Up to 250	>1.5	-	[142,143]
Trimethylene carbonate	−17.9 to −14.9	3.7–10.5 depending on mol. wt.	479 ± 165 to 622 ± 26	>24 *in vitro* [144] ~0.75 *in vivo* [145]	Yes [146]	[146,147]
Trimethylene carbonate-DLLA 50:50	11		570	<11 in vivo	Yes [146]	[97,148,149]
Trimethylene carbonate-DLLA 20:80	33	51	7	<11 in vivo	Yes [146]	[97,148,149]
Trimethylene carbonate-caprolactone 10:90	>−17	23	—	>24 in vivo	Yes [146]	[97,148,149]
Poly(ester amide)s	Variable	Variable	Variable	Variable	Yes [150–152]	[153]

*Source: Adapted from Shi et al. [125]. Reproduced with permission.

degradation sites [160], making it easier for the embedded cells to attach to and degrade the gel. Functionalizing gelatin with methacrylate groups makes it light polymerizable into a hydrogel with tunable mechanical properties that remains gelled above physiological temperatures [161]. This makes GelMA a promising candidate for soft tissue regeneration.

GelMA can also be supplemented with other components such as hyaluronic acid (HA), which is widely distributed throughout the extracellular matrix (ECM) of all connective tissues in humans [162,163]. It also plays an important role in many biological processes such as nutrient diffusion, tissue hydration, and cellular differentiation [164]. It is also biocompatible and biodegradable [164], which makes it an ideal component in biomedical hydrogel systems. More importantly, implantation of hydrogel-based scaffolds into nude mice resulted in local differentiation of preadipocytes, suggesting that this approach is feasible at least in an animal model [165].

Such hydrogels with tunable mechanical properties can also be "printed" with the help of an additive manufacturing process such as fused depositional modeling to create a hybrid scaffold containing a stiffer scaffold to maintain the mechanical properties of the construct, whereas the cell-laden hydrogels mimic the ECM of the tissue and provide a matrix that the cells can invade and remodel. Indeed, many attempts at bioprinting of hydrogels have already been successful [166–168]. Such hydrogel printing processes are also capable of depositing multiple cell types and bioactive factors such as proteins in controlled amounts to form cell-embedded tissue constructs [169]. Figure 10.10 shows the conceptual diagram of one such bioprinting system that can be used for breast tissue engineering applications.

10.3.2.1.3 Scaffold degradation
Degradation and erosion behaviors are also very important aspects of a biomaterial. Ideally, the degradation rate should match that of tissue regeneration, since scaffold erosion leads to an increase in the pore sizes, which facilitates the diffusion of molecules secreted by the cells (Figure 10.11) [170]. In addition, the scaffold degradation rate has been shown to influence cellular events such as cell spreading [171], proliferation [103], and protein expression [106].

10.3.2.1.4 Pore architecture
A high porosity and a high interconnectivity of the scaffold are desired to maximize the specific surface area for cell attachment and tissue ingrowth [172]. The pore morphology can also affect the growth of the cells and alter cell function [173]. Interconnected pores larger than the dimensions of the cells facilitate the infiltration of cells into the scaffold, whereas smaller pores positively influence the exchange of nutrients and cellular waste products [174].

10.3.2.2 Cells
A scaffold cannot initiate adipogenesis on its own. It needs to be seeded with appropriate cells for adipogenesis to occur. In case of adipose tissue engineering,

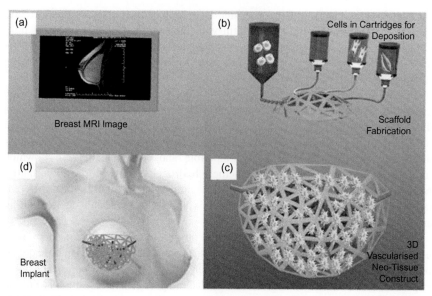

FIGURE 10.10

Conceptual diagram of a bioprinting system adapted for breast tissue reconstruction. A CAD model of the breast is constructed from an MRI or CT scan of the breast region (A), which is used to create a personalized scaffold containing cells and bioactive factors embedded in hydrogels (B). In vitro culturing of the tissue construct leads to preliminary vascularization of the construct (C), which can then be implanted back into the patient (D) for complete breast tissue reconstruction.

Source: Image taken from Melchels et al. [74]. Reproduced with permission.

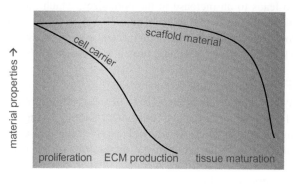

FIGURE 10.11

Graph showing the degradation of the scaffold over time interlayed with different cellular events taking place during tissue regeneration.

the most logical choice of cells would seem to be matured adipose cells. However, Patrick noted that adipose cells frequently form cysts or localized necrosis after injection [6]. Moreover, he also noted that they cannot be expanded in vitro because they are terminally differentiated—that is, they no longer proliferate. Adipose cells are also easily traumatized by the mechanical forces of liposuction due to the large volume of lipid content within them [54].

On the other hand, hAPCs, as defined previously, can be easily expanded in cell culture, are mechanically stable, and can easily be obtained from liposuction [175−178]. Moreover, they can easily be induced to differentiate into adipose tissue. These reasons establish hAPCs as the ideal cell type for seeding onto the scaffolds.

These hAPCs, or preadipocytes, also need a carrier to enhance attachment to the scaffold. Fibrin glue has been proposed as one such carrier by a number of studies [179−185]. In this case, cells are generally diluted in thrombin solution, which is later mixed with fibrinogen to create a fibrin matrix [118]. This matrix is then seeded onto the scaffold. Long-term stable adipose tissue formation with no inflammatory response in athymic nude mice has been successfully demonstrated using this strategy [185].

10.3.2.3 Microenvironment

In addition to having a cell-seeded scaffold, it is also important to create a microenvironment that supports the growth of the engineered tissue. Adipose tissue has been reported to be highly vascular [186]. In fact, Patrick [54] reports that each adipocyte is connected to at least one capillary, with the volume of capillary beds in adipose tissue being higher than even muscular tissue. Furthermore, hAPC cells die within 3 hours when deprived of oxygen [54]. This occurs because the diffusion of oxygen from the surrounding tissues is sufficient only over a distance of up to approximately 150 μm for adipose tissue [187]. Therefore, in tissue constructs without adequate vascularization, autologous adipose tissue loses up to 60 percent of its volume posttransplantation because the cells in the center of the graft are inadequately supplied with oxygen [188,189]. Hence, it becomes important to ensure that the tissue-engineered constructs have access to an adequate vascular network.

Some of the experimental approaches to rebuilding the native vascularization of tissue include endothelial cell induction, arteriovenous shunt loops, and vessel-embedded hydrogels.

10.3.2.3.1 Endothelial cell induction

This method, described by Borges et al. [185], involves the coimplantation of preadipocytes and endothelial cells. The rationale is that the endothelial cells would be able to differentiate relatively quickly to form numerous capillaries, thus supplying the preadipocytes with adequate oxygen supply. These capillaries then could connect to other capillaries on the periphery that have been induced by the host cells. This could then provide complete vascularization within the implant.

This concept of endothelial cell induction was developed further by Mertsching et al. [190], who developed methods to decellularize porcine small bowel segments and repopulated the remaining vascular structures with porcine endothelial progenitor cells. The authors suggest that such a vascularized scaffold could serve as a "universal scaffold" for tissue engineering.

A variation of this approach involves the stimulation of angiogenesis within a fibrin matrix by preadipocytes using growth factors such as vascular endothelial growth factor (VEGF), which stimulates vasculogenesis and angiogenesis, and basic fibroblast growth factor (bFGF), which mediates the formation of new blood vessels [191−193]. A recent study done by Ekaputra et al. [194] also indicated that the loading of VEGF and platelet-derived growth factor (PDGF) into a hybrid PCL/hydrogel scaffold increases its potential for complete vascularization.

10.3.2.3.2 Bioreactors

Adipose cells require substantial vascularization and efficient exchange of nutrients in order to proliferate. The simple loading of cells to scaffolds has been limited by the challenge of maintaining the cellular viability at the core [195]. Therefore, bioreactors are used in tissue engineering to improve the cell density and homogeneity on the surface of the scaffold [196−198]. Bioreactors are essentially mechanical devices that create controllable, mechanically active environments to aid in the effective seeding of cells onto a polymer [199]. Research has also shown that larger constructs with improved cellularity upon cultivation have been obtained with the help of bioreactors [200,201].

One such bioreactor that holds much promise in the case of breast tissue engineering is the biaxial bioreactor developed by the National University of Singapore, shown in Figure 10.12.

The bioreactor rotates simultaneously in two independent axes. This spinning movement leads to an improved flow of the fluids within the bioreactor, as compared to static cell culture [195]. Experiments have shown that this bioreactor allowed the maintenance of cellular viability at the core beyond the limits of conventional diffusion, with increased cell proliferation and differentiation [195].

10.3.2.3.3 Cell-embedded hydrogels for bioprinting

This is an emerging approach to vascularization developed by Chrobak et al. [202]. First of all, approximately 100-μm-diameter open channels are etched, using lithography techniques, on a collagen hydrogel. Endothelial cells are then cultured on the surface of the gel. Because these channels are etched, perfusion is readily established upon seeding. This leads to a formation of microvascular tubes of endothelial cells with diameters ranging from 75 to 150 μm. A similar approach could also be investigated in breast tissue engineering: fabricating a hybrid polymer + hAPC and endothelial cell-embedded hydrogel scaffold to promote vascularization.

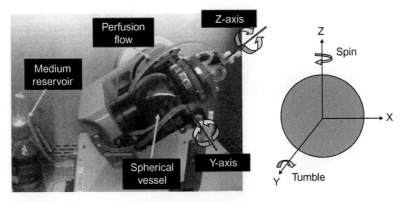

FIGURE 10.12

Biaxial bioreactor design. The bioreactor system consists of a spherical culture vessel (volume of 500 mL) connected to the medium reservoir through tubing through which a perfusion flow is generated (flow direction indicated by red arrows). The spherical vessel sits on an articulator that allows rotation in two perpendicular axes (X and Z).

Source: Zhang et al. [195]. Reproduced with permission.

10.3.2.3.4 Arteriovenous shunt loops

An arteriovenous shunt (or AV shunt) is a passageway, usually a U-shaped tube, between an artery and a vein. The blood then passes directly from an artery into a vein, bypassing the capillaries. This graft is then implanted into the test subject in vivo. Erol and Spira [203] first reported the formation of a new capillary bed around such an implanted AV shunt loop. Subsequent studies done by Tanaka et al. [204], Mian et al. [205], and Lokmic et al. [206] showed that AV shunt loops cannot only induce de novo vascularization but also the formation of new tissue. Furthermore, it has been reported by Wiggenhauser et al. [118] that plastic surgeons are familiar with the surgery of small vessels, so this method can easily be transferred to the clinic. Figure 10.13 shows an AV shunt loop.

CONCLUSION

In conclusion, medical imaging and additive manufacturing have the potential to improve the outcome of breast reconstruction. We foresee that AM technologies will be further developed and tailored to this application and on the longer term will be introduced into the clinic in different stages with increasing levels of complexity.

First, in the very near future, AM technologies can be employed to create molds that aid the surgeon in free tissue transfer flap surgeries. This approach poses no strict requirements on the technique or materials used and has low risk, but it can significantly improve the aesthetic outcome. In a second stage,

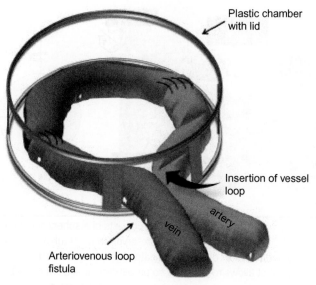

Plastic chamber with lid

Insertion of vessel loop

artery

vein

Arteriovenous loop fistula

FIGURE 10.13

An AV shunt loop model. An artery is anastomosed to a vein graft and then to a vein to form the vascular loop. This loop is then placed inside a plastic chamber with an open end.

Source: From [207]. Reproduced with permission.

personalized implants could be manufactured again to improve the outcome of the acellular (silicon) implant approach. Third, personalized degradable scaffolds could be produced that, after seeding with co-cultures of relevant cell types, can be induced to form vascularized fat tissue for breast reconstruction. The final milestone would be the production of personalized, cell-containing, fully organized 3D tissue constructs in a computer-controlled manner that would enable full regeneration of the native breast.

Although rapid progress is being made in this area, the technology can still be considered to be in its infancy. Future efforts need to be directed in terms of the integration of different additive manufacturing processes to include cells, biomaterials, growth factors, and extracellular matrix components into an integrated system. It is also important to find suitable biomaterials and hydrogels that match the rheological and mechanical properties of the breast environment and tissue degradation rate. Such an integrated approach would require an interdisciplinary team combining expertise in polymer chemistry, computer engineering, mechanical engineering, biology, and medicine. With such a joint effort, biofabrication-based tissue engineering strategies have the potential to evolve into a technology platform that allows surgeons and scientists alike to create tissue engineered constructs for regeneration of many tissues including for breast reconstruction.

References

[1]   Bray F, McCarron P, Parkin DM. The changing global patterns of female breast cancer incidence and mortality. Breast Cancer Res 2004;4:5.

[2]   Renneker R, Cutler M. Psychological problems of adjustment to cancer of the breast. JAMA 1952;148(10):833.

[3]   Surgeons ASoP. Cosmetic procedure trends. Plastic surgery statistics report, American Society of Plastic Surgeons, 2011.

[4]   Cooper SA. On the anatomy of the breast. London: Longman, Orme, Green, Brown, and Longmans; 1840.

[5]   Moore KL, Dalley AF, Agur AMR. Clinically oriented anatomy, 5. Philadelphia, PA: Lippincott Williams & Wilkins; 1999.

[6]   Patrick CW. Breast tissue engineering. Annu Rev Biomed Eng 2004;6(1):109−30.

[7]   Moore K. Clinically oriented anatomy. Baltimore: Williams & Wilkins; 1992.

[8]   Ramsay D, et al. Anatomy of the lactating human breast redefined with ultrasound imaging. J Anat 2005;206(6):525−34.

[9]   Robb GL. Reconstructive surgery. In: Hunt K, Robb G, Strom E, editors. Breast cancer. New York: Springer-Verlag; 2001. p. 223−53.

[10] Muzaffar AR, Rohrich RJ. The silicone gel-filled breast implant controversy: an update. Plast Reconstr Surg 2002;109(2):742.

[11] Flassbeck D, et al. Determination of siloxanes, silicon, and platinum in tissues of women with silicone gel-filled implants. Anal Bioanal Chem 2003;375(3):356−62.

[12] Pollock H. Breast capsular contracture: a retrospective study of textured versus smooth silicone implants. Plast Reconstr Surg 1993;91:404.

[13] Peters W, et al. Capsular calcification associated with silicone breast implants: incidence, determinants, and characterization. Ann Plast Surg 1998;41(4):348.

[14] Coleman DJ, Foo ITH, Sharpe DT. Textured or smooth implants for breast augmentation? A prospective controlled trial. Br J Plast Surg 1991;44(6):444−8.

[15] Baran CN, et al. A different strategy in the surgical treatment of capsular contracture: leave capsule intact. Aesthetic Plast Surg 2001;25(6):427−31.

[16] Gerszten PC. A formal risk assessment of silicone breast implants. Biomaterials 1999;20(11):1063−9.

[17] Burkhardt B, et al. Capsular contracture: a prospective study of the effect of local antibacterial agents. Plast Reconstr Surg 1986;77(6):919.

[18] Asplund O. Capsular contracture in silicone gel and saline-filled breast implants after reconstruction. Plast Reconstr Surg 1984;73(2):270.

[19] Hakelius L, Ohlsén L. A clinical comparison of the tendency to capsular contracture between smooth and textured gel-filled silicone mammary implants. Plast Reconstr Surg 1992;90(2):247.

[20] Gylbert L, Asplund O, Jurell G. Capsular contracture after breast reconstruction with silicone-gel and saline-filled implants: a 6-year follow-up. Plast Reconstr Surg 1990;85(3):373.

[21] Contant CME, et al. Morbidity of immediate breast reconstruction (IBR) after mastectomy by a subpectorally placed silicone prosthesis: the adverse effect of radiotherapy. Eur J Surg Oncol (EJSO) 2000;26(4):344−50.

[22] Vandeweyer E, Deraemaecker R. Radiation therapy after immediate breast reconstruction with implants. Plast Reconstr Surg 2000;106(1):56.

[23] Jacobson GM, et al. Breast irradiation following silicone gel implants. Int J Radiat Oncol Biol Phys 1986;12(5):835–8.

[24] Rosato RM, Dowden RV. Radiation therapy as a cause of capsular contracture. Ann Plast Surg 1994;32(4):342.

[25] Ma PX, Elisseeff J. Scaffolding in tissue engineering. Boca Raton, Florida, USA: CRC; 2005.

[26] Clough KB, et al. Prospective evaluation of late cosmetic results following breast reconstruction: I. Implant reconstruction. Plast Reconstr Surg 2001;107(7):1702.

[27] Cordeiro PG. Breast reconstruction after surgery for breast cancer. N Engl J Med 2008;359(15):1590–601.

[28] Rozen WM, et al. Post-mastectomy breast reconstruction: a history in evolution. Clin Breast Cancer 2009;9(3):145–54.

[29] Niechajev I, Sevćuk O. Long-term results of fat transplantation: clinical and histologic studies. Plast Reconstr Surg 1994;94(3):496.

[30] Matsudo PKR, Toledo LS. Experience of injected fat grafting. Aesthetic Plast Surg 1988;12(1):35–8.

[31] Lee K, et al. Breast reconstruction. In: Lanza R, Vacanti J, editors. Principle of tissue engineering. San Diego: Academic Press; 2000.

[32] Khouri R. Breast reconstruction after mastectomy. Available from: <http://www.miamibreastcenter.com/reconstruction/after-mastectomy-miami>; 2011.

[33] Coleman SR. Facial recontouring with lipostructure. Clin Plast Surg 1997;24(2):347.

[34] Hartrampf C, Scheflan M, Black PW. Breast reconstruction with a transverse abdominal island flap. Plast Reconstr Surg 1982;69(2):216.

[35] Allen RJ, Treece P. Deep inferior epigastric perforator flap for breast reconstruction. Ann Plast Surg 1994;32(1):32.

[36] Gill PS, et al. A 10-year retrospective review of 758 DIEP flaps for breast reconstruction. Plast Reconstr Surg 2004;113(4):1153.

[37] Blondeel PN, et al. Venous congestion and blood flow in free transverse rectus abdominis myocutaneous and deep inferior epigastric perforator flaps. Plast Reconstr Surg 2000;106(6):1295.

[38] Nahabedian MY, et al. Breast reconstruction with the free TRAM or DIEP flap: patient selection, choice of flap, and outcome. Plast Reconstr Surg 2002;110(2):466.

[39] Feller AM. Free TRAM. Results and abdominal wall function. Clin Plast Surg 1994;21(2):223.

[40] Arnez Z, et al. Rational selection of flaps from the abdomen in breast reconstruction to reduce donor site morbidity. Br J Plast Surg 1999;52(5):351–4.

[41] Serletti JM, Moran SL. Free versus the pedicled TRAM flap: a cost comparison and outcome analysis. Plast Reconstr Surg 1997;100(6):1418.

[42] Serletti JM, Moran SL. Microvascular reconstruction of the breast. Wiley Online Library; 2000.

[43] Kroll SS, et al. Fat necrosis in free and pedicled TRAM flaps. Plast Reconstr Surg 1998;102(5):1502.

[44] Kroll SS, et al. Abdominal wall strength, bulging, and hernia after TRAM flap breast reconstruction. Plast Reconstr Surg 1995;96(3):616.

[45] Kroll SS, Marchi M. Comparison of strategies for preventing abdominal-wall weakness after TRAM flap breast reconstruction. Plast Reconstr Surg 1992;89(6):1045.

[46] Edsander-Nord Å, Jurell G, Wickman M. Donor-site morbidity after pedicled or free TRAM flap surgery: a prospective and objective study. Plast Reconstr Surg 1998;102 (5):1508.

[47] Blondeel P. One hundred free DIEP flap breast reconstructions: a personal experience. Br J Plast Surg 1999;52(2):104−11.

[48] Kroll SS. Fat necrosis in free transverse rectus abdominis myocutaneous and deep inferior epigastric perforator flaps. Plast Reconstr Surg 2000;106(3):576.

[49] Patrick Jr C, et al. Long-term implantation of preadipocyte-seeded PLGA scaffolds. Tissue Eng 2002;8(2):283−93.

[50] Patrick Jr C, et al. Preadipocyte seeded PLGA scaffolds for adipose tissue engineering. Tissue Eng 1999;5(2):139−51.

[51] Entenmann G, Hauner H. Relationship between replication and differentiation in cultured human adipocyte precursor cells. Am J Physiol Cell Physiol 1996;270(4): C1011−6.

[52] Hausman G, Richardson R. Newly recruited and pre-existing preadipocytes in cultures of porcine stromal-vascular cells: morphology, expression of extracellular matrix components, and lipid accretion. J Anim Sci 1998;76(1):48.

[53] Shillabeer G, Forden J, Lau D. Induction of preadipocyte differentiation by mature fat cells in the rat. J Clin Invest 1989;84(2):381.

[54] Patrick Jr CW. Adipose tissue engineering: the future of breast and soft tissue reconstruction following tumor resection. Wiley Online Library; 2000.

[55] Park J, DeSouza GN, Kak. AC. Dual-beam structured-light scanning for 3-D object modeling. IEEE; 2001.

[56] Kovacs L, et al. New aspects of breast volume measurement using 3-dimensional surface imaging. Ann Plast Surg 2006;57(6):602.

[57] Kovacs L, et al. Comparison between breast volume measurement using 3D surface imaging and classical techniques. Breast 2007;16(2):137−45.

[58] Eder M, et al. [Breast volume assessment based on 3D surface geometry: verification of the method using MR imaging]. Biomed Tech Biomed Eng 2008;53(3):112.

[59] Edinger M, et al. Advancing animal models of neoplasia through *in vivo* bioluminescence imaging. Eur J Cancer 2002;38(16):2128−36.

[60] Contag CH, et al. Use of reporter genes for optical measurements of neoplastic disease in vivo. Neoplasia (New York, NY) 2000;2(1-2):41.

[61] Edinger M, et al. Noninvasive assessment of tumor cell proliferation in animal models. Neoplasia (New York, NY) 1999;1(4):303.

[62] Sweeney TJ, et al. Visualizing the kinetics of tumor-cell clearance in living animals. Proc Natl Acad Sci USA 1999;96(21):12044.

[63] Contag CH, et al. Photonic detection of bacterial pathogens in living hosts. Mol Microbiol 1995;18(4):593−603.

[64] Beekman FJ, et al. Towards in vivo nuclear microscopy: iodine-125 imaging in mice using micro-pinholes. Eur J Nucl Med Mol Imaging 2002;29(7):933−8.

[65] Frankle WG, et al. Neuroreceptor imaging in psychiatry: theory and applications. Int Rev Neurobiol 2005;67:385−440.

[66] Herman GT. Fundamentals of computerized tomography: image reconstruction from projections. Springer Verlag; 2009.

[67] Weber DA, Ivanovic M. Ultra-high-resolution imaging of small animals: implications for preclinical and research studies. J Nucl Cardiol 1999;6(3):332−44.

[68] Hendee WR, Morgan CJ. Magnetic resonance imaging part I—physical principles. West J Med 1984;141(4):491.

[69] Hohe J, et al. Surface size, curvature analysis, and assessment of knee joint incongruity with MRI in vivo. Magn Reson Med 2002;47(3):554—61.

[70] Smith-Bindman R, et al. Radiation dose associated with common computed tomography examinations and the associated lifetime attributable risk of cancer. Arch Intern Med 2009;169(22):2078.

[71] Berrington de Gonzalez A, et al. Projected cancer risks from computed tomographic scans performed in the United States in 2007. Arch Intern Med 2009;169(22):2071.

[72] Melchels F, et al. CAD/CAM-assisted breast reconstruction. Biofabrication 2011;3:034114.

[73] Chua C, et al. An integrated experimental approach to link a laser digitiser, a CAD/CAM system and a rapid prototyping system for biomedical applications. Int J Adv Manuf Technol 1998;14(2):110—5.

[74] Melchels FPW, et al. Additive manufacturing of tissues and organs. Prog Polym Sci 2011.

[75] Sachlos E, Czernuszka J. Making tissue engineering scaffolds work. Review: the application of solid freeform fabrication technology to the production of tissue engineering scaffolds. Eur Cell Mater 2003;5(29):39—40.

[76] Tan Q, et al. Fabrication of porous scaffolds with a controllable microstructure and mechanical properties by porogen fusion technique. Int J Mol Sci 2011;12(2):890—904.

[77] Hutmacher DW. Scaffold design and fabrication technologies for engineering tissues: state of the art and future perspectives. J Biomater Sci Polym Ed 2001;12(1):107—24.

[78] Zein I, et al. Fused deposition modeling of novel scaffold architectures for tissue engineering applications. Biomaterials 2002;23(4):1169—85.

[79] Hutmacher DW, Sittinger M, Risbud MV. Scaffold-based tissue engineering: rationale for computer-aided design and solid free-form fabrication systems. Trends Biotechnol 2004;22(7):354—62.

[80] Hutmacher DW, et al. Mechanical properties and cell cultural response of polycaprolactone scaffolds designed and fabricated via fused deposition modeling. J Biomed Mater Res 2001;55(2):203—16.

[81] Domingos M, et al. Polycaprolactone scaffolds fabricated via bioextrusion for tissue engineering applications. Int J Biomater 2009.

[82] Melchels FPW. Preparation of advanced porous structures by stereolithography for application in tissue engineering. Netherlands: University of Twente; 2010.

[83] Xu T, et al. Inkjet printing of viable mammalian cells. Biomaterials 2005;26(1):93—9.

[84] Mironov V, et al. Organ printing: computer-aided jet-based 3D tissue engineering. TRENDS Biotechnol 2003;21(4):157—61.

[85] Williams JM, et al. Bone tissue engineering using polycaprolactone scaffolds fabricated via selective laser sintering. Biomaterials 2005;26(23):4817—27.

[86] Cheung HY, et al. A critical review on polymer-based bio-engineered materials for scaffold development. Composites Part B: Eng 2007;38(3):291—300.

[87] Patrick CW, Mikos AG, McIntire LV. Frontiers in tissue engineering. Oxford: Permagon; 1998.

[88]   Marler JJ, et al. Soft-tissue augmentation with injectable alginate and syngeneic fibroblasts. Plast Reconstr Surg 2000;105(6):2049.

[89]   Lemperle G, Morhenn V, Charrier U. Human histology and persistence of various injectable filler substances for soft tissue augmentation. Aesthetic Plast Surg 2003;27(5):354−66.

[90]   Ducheyne P, Healy KE, et al., editors. Comprehensive biomaterials. Oxford, UK: Elsevier; 2011.

[91]   Ceballos D, et al. Magnetically aligned collagen gel filling a collagen nerve guide improves peripheral nerve regeneration. Exp Neurol 1999;158(2):290−300.

[92]   Compton CC, et al. Organized skin structure is regenerated in vivo from collagen-GAG matrices seeded with autologous keratinocytes. J Invest Dermatol 1998;110(6):908−16.

[93]   Mooney DJ, et al. Long-term engraftment of hepatocytes transplanted on biodegradable polymer sponges. J Biomed Mater Res 1997;37(3):413−20.

[94]   Vacanti JP, et al. Selective cell transplantation using bioabsorbable artificial polymers as matrices*. J Pediatr Surg 1988;23(1):3−9.

[95]   Niklason L, et al. Functional arteries grown in vitro. Science 1999;284(5413):489−93.

[96]   Matsumoto K, et al. Use of a newly developed artificial nerve conduit to assist peripheral nerve regeneration across a long gap in dogs. ASAIO J 2000;46(4):415.

[97]   Pêgo AP, et al. Biodegradable elastomeric scaffolds for soft tissue engineering. J Control Release 2003;87(1):69−79.

[98]   Engler AJ, et al. Matrix elasticity directs stem cell lineage specification. Cell 2006;126(4):677−89.

[99]   McBride SH, Falls T, Tate MLKnothe. Modulation of stem cell shape and fate B: mechanical modulation of cell shape and gene expression. Tissue Eng Part A 2008;14(9):1573−80.

[100] Tanzi MC, Farè S. Adipose tissue engineering: state of the art, recent advances and innovative approaches. Expert Rev Med Devices 2009;6(5):533−51.

[101] Pelham RJ, Wang Y. Cell locomotion and focal adhesions are regulated by substrate flexibility. Proc Natl Acad Sci USA 1997;94(25):13661.

[102] Rowley JA, Mooney DJ. Alginate type and RGD density control myoblast phenotype. J Biomed Mater Res 2002;60(2):217−23.

[103] Boontheekul T, et al. Regulating myoblast phenotype through controlled gel stiffness and degradation. Tissue Eng 2007;13(7):1431−42.

[104] Gray DS, Tien J, Chen CS. Repositioning of cells by mechanotaxis on surfaces with micropatterned Young's modulus. J Biomed Mater Res Part A 2003;66 (3):605−14.

[105] Peyton SR, et al. The effects of matrix stiffness and RhoA on the phenotypic plasticity of smooth muscle cells in a 3-D biosynthetic hydrogel system. Biomaterials 2008;29(17):2597−607.

[106] Bryant SJ, Anseth KS. Hydrogel properties influence ECM production by chondrocytes photoencapsulated in poly (ethylene glycol) hydrogels. J Biomed Mater Res 2002;59(1):63−72.

[107] Bryant SJ, et al. Crosslinking density influences chondrocyte metabolism in dynamically loaded photocrosslinked poly (ethylene glycol) hydrogels. Ann Biomed Eng 2004;32(3):407−17.

[108] Grijpma DW, et al. Improvement of the mechanical properties of poly (D, L-lactide) by orientation. Polym Int 2002;51(10):845−51.

[109] Hiljanen-Vainio M, Karjalainen T, Seppälä J. Biodegradable lactone copolymers. I. Characterization and mechanical behavior of ε-caprolactone and lactide copolymers. J Appl Polym Sci 1996;59(8):1281–8.

[110] Gay S, Arostegui S, Lemaitre J. Preparation and characterization of dense nanohydroxyapatite/PLLA composites. Mater Sci Eng: C 2009;29(1):172–7.

[111] Papenburg BJ. Design strategies for tissue engineering scaffolds. Netherlands: University of Twente; 2009, PhD Thesis.

[112] Williamson MR, Adams EF, Coombes AGA. Cell attachment and proliferation on novel polycaprolactone fibres having application in soft tissue engineering. Eur Cell Mater 2002;4(2):62–3.

[113] Gefen A, Dilmoney B. Mechanics of the normal woman's breast. Technol Health Care 2007;15(4):259–71.

[114] Wiggenhauser PS, et al. Engineering of vascularized adipose constructs. Cell Tissue Res 2011;347(3):747–57.

[115] Woodruff MA, Hutmacher DW. The return of a forgotten polymer—polycaprolactone in the 21st century. Prog Polym Sci 2010;35(10):1217–56.

[116] Lam CXF, et al. Evaluation of polycaprolactone scaffold degradation for 6 months in vitro and in vivo. J Biomed Mater Res Part A 2009;90(3):906–19.

[117] Weiser B, et al. In vivo development and long-term survival of engineered adipose tissue depend on in vitro precultivation strategy. Tissue Eng Part A 2008;14(2):275–84.

[118] Wiggenhauser PS, et al. Engineering of vascularized adipose constructs. Cell Tissue Res 2011;347(3):747–57.

[119] Cheng Z, Teoh SH. Surface modification of ultra thin poly (ε-caprolactone) films using acrylic acid and collagen. Biomaterials 2004;25(11):1991–2001.

[120] Zhu Y, et al. Surface modification of polycaprolactone membrane via aminolysis and biomacromolecule immobilization for promoting cytocompatibility of human endothelial cells. Biomacromolecules 2002;3(6):1312–9.

[121] Zhu Y, Gao C, Shen J. Surface modification of polycaprolactone with poly (methacrylic acid) and gelatin covalent immobilization for promoting its cytocompatibility. Biomaterials 2002;23(24):4889–95.

[122] Cowie JMG, Arrighi V. Polymers: chemistry and physics of modern materials. J Polym Sci, Part A: Polym Chem 1992;30(8):1777.

[123] Amsden B. Curable, biodegradable elastomers: emerging biomaterials for drug delivery and tissue engineering. Soft Matter 2007;3(11):1335–48.

[124] Serrano MC, Chung EJ, Ameer GA. Advances and applications of biodegradable elastomers in regenerative medicine. Adv Funct Mater 2010;20(2):192–208.

[125] Shi R, et al. Recent advances in synthetic bioelastomers. Int J Mol Sci 2009;10 (10):4223–56.

[126] Whatley BR, et al. Fabrication of a biomimetic elastic intervertebral disk scaffold using additive manufacturing. Biofabrication 2011;3:015004.

[127] Agrawal CM, Parr JE, Lin ST. Synthetic bioabsorbable polymers for implants. Astm Intl.; 2000.

[128] Tatai L, et al. Thermoplastic biodegradable polyurethanes: the effect of chain extender structure on properties and in-vitro degradation. Biomaterials 2007;28(36):5407–17.

[129] Bhattacharyya S, et al. Development of biodegradable polyphosphazene-nanohydroxyapatite composite nanofibers via electrospinning. Cambridge Univ Press; 2004.

[130] Schacht E, et al. Biomedical applications of degradable polyphosphazenes. Biotechnol Bioeng 1996;52(1):102−8.

[131] Honarkar H, Rahimi A. Applications of inorganic polymeric materials, III: polyphosphazenes. MonatshChem/Chem Mon 2007;138(10):923−33.

[132] Yi F, LaVan DA. Poly (glycerol sebacate) nanofiber scaffolds by core/shell electrospinning. Macromol Biosci 2008;8(9):803−6.

[133] Wang Y, et al. A tough biodegradable elastomer. Nat Biotechnol 2002;20 (6):602−6.

[134] Wang Y, Kim YM, Langer R. In vivo degradation characteristics of poly (glycerol sebacate). J Biomed Mater Res Part A 2003;66(1):192−7.

[135] Jeong CG, Hollister SJ. Mechanical, permeability, and degradation properties of 3D designed poly (1, 8 octanediol-co-citrate) scaffolds for soft tissue engineering. J Biomed Mater Res Part B: Appl Biomater 2010;93(1):141−9.

[136] Yang J, Webb AR, Ameer GA. Novel citric acid-based biodegradable elastomers for tissue engineering. Adv Mater 2004;16(6):511−6.

[137] Yang J, et al. Synthesis and evaluation of poly (diol citrate) biodegradable elastomers. Biomaterials 2006;27(9):1889−98.

[138] Moroni L, et al. Fiber diameter and texture of electrospun PEOT/PBT scaffolds influence human mesenchymal stem cell proliferation and morphology, and the release of incorporated compounds. Biomaterials 2006;27(28):4911−22.

[139] Catalani LH, Collins G, Jaffe M. Evidence for molecular orientation and residual charge in the electrospinning of poly (butylene terephthalate) nanofibers. Macromolecules 2007;40(5):1693−7.

[140] Mathew G, et al. Preparation and characterization of properties of electrospun poly (butylene terephthalate) nanofibers filled with carbon nanotubes. Polym Test 2005;24(6):712−7.

[141] Deschamps AA, et al. Design of segmented poly (ether ester) materials and structures for the tissue engineering of bone. J Control Release 2002;78(1):175−86.

[142] Webb AR, Yang J, Ameer GA. Biodegradable polyester elastomers in tissue engineering. Expert Opin Biol Ther 2004;4(6):801−12.

[143] Lee SH, et al. Elastic biodegradable poly (glycolide-co-caprolactone) scaffold for tissue engineering. J Biomed Mater Res Part A 2003;66(1):29−37.

[144] Pego AP, et al. In vitro degradation of trimethylene carbonate based (co) polymers. Macromol Biosci 2002;2(9):411−9.

[145] Zhang Z, et al. The in vivo and in vitro degradation behavior of poly (trimethylene carbonate). Biomaterials 2006;27(9):1741−8.

[146] Bat E, et al. Ultraviolet light crosslinking of poly (trimethylene carbonate) for elastomeric tissue engineering scaffolds. Biomaterials 2010;31(33):8696−705.

[147] Bat E, et al. Crosslinking of trimethylene carbonate and D, L-lactide (co-) polymers by gamma irradiation in the presence of tentaerythritol triacrylate. Macromol Biosci 2011;11(7):952−61.

[148] Pego A, et al. In vivo behavior of poly (1, 3-trimethylene carbonate) and copolymers of 1, 3-trimethylene carbonate with D, L-lactide or -caprolactone: degradation and tissue response. J Biomed Mater Res Part A 2003;67(3):1044−54.

[149] Pego AP, et al. Physical properties of high molecular weight 1, 3-trimethylene carbonate and D, L-lactide copolymers. J Mater Sci Mater Med 2003;14(9): 767−73.

[150] Dong Y, et al. Degradation behaviors of electrospun resorbable polyester nanofibers. Tissue Eng Part B: Rev 2009;15(3):333–51.

[151] Li L, Chu CC. Nitroxyl radical incorporated electrospun biodegradable poly (ester amide) nanofiber membranes. J Biomater Sci Polym Ed 2009;20(3):341–61.

[152] Garg P, et al. Electrospinning of novel poly (ester amide)s. Macromol Mater Eng 2009;294(10):679–90.

[153] Tsitlanadze G, et al. In vitro enzymatic biodegradation of amino acid based poly (ester amide)s biomaterials. J Mater Sci Mater Med 2004;15(2):185–90.

[154] Nichol JW, et al. Cell-laden microengineered gelatin methacrylate hydrogels. Biomaterials 2010;31(21):5536–44.

[155] Du Y, et al. Directed assembly of cell-laden microgels for fabrication of 3D tissue constructs. Proc Natl Acad Sci USA 2008;105(28):9522.

[156] Khademhosseini A, et al. Micromolding of photocrosslinkable hyaluronic acid for cell encapsulation and entrapment. J Biomed Mater Res Part A 2006;79(3):522–32.

[157] Brigham MD, et al. Mechanically robust and bioadhesive collagen and photocross-linkable hyaluronic acid semi-interpenetrating networks. Tissue Eng Part A 2008;15 (7):1645–53.

[158] Van Den Bulcke AI, et al. Structural and rheological properties of methacrylamide modified gelatin hydrogels. Biomacromolecules 2000;1(1):31–8.

[159] Galis ZS, Khatri JJ. Matrix metalloproteinases in vascular remodeling and atherogenesis. Circ Res 2002;90(3):251–62.

[160] Van den Steen PE, et al. Biochemistry and molecular biology of gelatinase B or matrix metalloproteinase-9 (MMP-9). Crit Rev Biochem Mol Biol 2002;37(6):375–536.

[161] Benton JA, et al. Photocrosslinking of gelatin macromers to synthesize porous hydrogels that promote valvular interstitial cell function. Tissue Eng Part A 2009;15 (11):3221–30.

[162] Goa KL, Benfield P. Hyaluronic acid. A review of its pharmacology and use as a surgical aid in ophthalmology, and its therapeutic potential in joint disease and wound healing. Drugs 1994;47(3):536.

[163] Toole B. Hyaluronan and its binding proteins, the hyaladherins. Curr Opin Cell Biol 1990;2(5):839.

[164] Tan H, et al. Thermosensitive injectable hyaluronic acid hydrogel for adipose tissue engineering. Biomaterials 2009;30(36):6844–53.

[165] von Heimburg D, et al. Influence of different biodegradable carriers on the in vivo behavior of human adipose precursor cells. Plast Reconstr Surg 2001;108(2):411.

[166] Song S-J, et al. Sodium alginate hydrogel-based bioprinting using a novel multinozzle bioprinting system. Artif Organs 2011;35(11):1132–6.

[167] Pescosolido L, et al. Hyaluronic acid and dextran based semi-IPN hydrogels as biomaterials for bioprinting. Biomacromolecules 2011.

[168] Chang R, Nam J, Sun W. Computer-aided design, modeling, and freeform fabrication of 3D tissue constructs for drug metabolism studies. Comput Aided Des Appl 2008;5:21–9.

[169] Khalil S, Sun W. Biopolymer deposition for freeform fabrication of hydrogel tissue constructs. Mater Sci Eng: C 2007;27(3):469–78.

[170] Bat E. Elastomeric networks based on trimethylene carbonate polymers for biomedical applications: physical properties and degradation behaviour. Netherlands: University of Twente; 2010, PhD Thesis.

[171] Hudalla GA, Eng TS, Murphy WL. An approach to modulate degradation and mesenchymal stem cell behavior in poly (ethylene glycol) networks. Biomacromolecules 2008;9(3):842–9.

[172] Hou Q, Grijpma DW, Feijen J. Preparation of interconnected highly porous polymeric structures by a replication and freeze-drying process. J Biomed Mater Res Part B Appl Biomater 2003;67(2):732–40.

[173] Schugens C, et al. Biodegradable and macroporous polylactide implants for cell transplantation: 1. Preparation of macroporous polylactide supports by solid-liquid phase separation. Polymer 1996;37(6):1027–38.

[174] Zhang R, Ma PX. Synthetic nano-fibrillar extracellular matrices with predesigned macroporous architectures. 2000.

[175] Patel P, Robb GL, Patrick Jr CW. Soft tissue restoration using tissue engineering. Semin Plast Surg 2003;17:99–106.

[176] Patrick Jr C. Tissue engineering strategies for adipose tissue repair. Anat Rec 2001;263(4):361–6.

[177] Beahm EK, Walton RL, Patrick Jr CW. Progress in adipose tissue construct development. Clin Plast Surg 2003;30(4):547.

[178] Patrick Jr CW, et al. Epithelial cell culture: breast. San Diego: Academic Press; 2001.

[179] Torio-Padron N, et al. Engineering of adipose tissue by injection of human preadipocytes in fibrin. Aesthetic plastic surgery 2007;31(3):285–93.

[180] Cronin KJ, et al. The role of biological extracellular matrix scaffolds in vascularized three-dimensional tissue growth in vivo. J Biomed Mater Res Part B: Appl Biomater 2007;82(1):122–8.

[181] Cho SW, et al. Engineering of volume-stable adipose tissues. Biomaterials 2005;26 (17):3577–85.

[182] Wechselberger G, et al. Successful transplantation of three tissue-engineered cell types using capsule induction technique and fibrin glue as a delivery vehicle. Plast Reconstr Surg 2002;110(1):123.

[183] Tabata Y, et al. De novo formation of adipose tissue by controlled release of basic fibroblast growth factor. Tissue Eng 2000;6(3):279–89.

[184] Ahmed TAE, Dare EV, Hincke M. Fibrin: a versatile scaffold for tissue engineering applications. Tissue Eng Part B: Rev 2008;14(2):199–215.

[185] Borges J, et al. Engineered adipose tissue supplied by functional microvessels. Tissue Eng 2003;9(6):1263–70.

[186] Crandall DL, Hausman GJ, Kral JG. A review of the microcirculation of adipose tissue: anatomic, metabolic, and angiogenic perspectives. Microcirculation 1997;4(2):211–32.

[187] Awwad HK, et al. Intercapillary distance measurement as an indicator of hypoxia in carcinoma of the cervix uteri. Int J Radiat Oncol Biol Phys 1986;12(8):1329–33.

[188] Smahel J. Experimental implantation of adipose tissue fragments. Br J Plast Surg 1989;42(2):207–11.

[189] Czerny M. Reconstruction of the breast with a lipoma. Chir Kongr Verh 1895;2:216.

[190] Mertsching H, et al. Engineering of a vascularized scaffold for artificial tissue and organ generation. Biomaterials 2005;26(33):6610–7.

[191] Wilting J, et al. In vivo effects of vascular endothelial growth factor on the chicken chorioallantoic membrane. Cell Tissue Res 1993;274(1):163–72.

[192] Oh SJ, et al. VEGF and VEGF-C: specific induction of angiogenesis and lymphangiogenesis in the differentiated avian chorioallantoic membrane. Dev Biol 1997;188 (1):96−109.

[193] Ribatti D, et al. Endogenous basic fibroblast growth factor is implicated in the vascularization of the chick embryo chorioallantoic membrane. Dev Biol 1995;170(1):39−49.

[194] Ekaputra AK, et al. The three-dimensional vascularization of growth factor-releasing hybrid scaffold of poly (-caprolactone)/collagen fibers and hyaluronic acid hydrogel. Biomaterials 2011.

[195] Zhang ZY, et al. A biaxial rotating bioreactor for the culture of fetal mesenchymal stem cells for bone tissue engineering. Biomaterials 2009;30(14):2694−704.

[196] Radisic M, et al. High-density seeding of myocyte cells for cardiac tissue engineering. Biotechnol Bioeng 2003;82(4):403−14.

[197] Radisic M, et al. Medium perfusion enables engineering of compact and contractile cardiac tissue. Am J Physiol Heart Circ Physiol 2004;286(2):H507−16.

[198] Bursac N, et al. Cultivation in rotating bioreactors promotes maintenance of cardiac myocyte electrophysiology and molecular properties. Tissue Eng 2003;9(6):1243−53.

[199] Freed LE, et al. Advanced tools for tissue engineering: scaffolds, bioreactors, and signaling. Tissue Eng 2006;12(12):3285−305.

[200] Martin I, Wendt D, Heberer M. The role of bioreactors in tissue engineering. TRENDS Biotechnol 2004;22(2):80−6.

[201] Pei M, et al. Bioreactors mediate the effectiveness of tissue engineering scaffolds. FASEB J 2002;16(12):1691−4.

[202] Chrobak KM, Potter DR, Tien J. Formation of perfused, functional microvascular tubes in vitro. Microvasc Res 2006;71(3):185−96.

[203] Erol OO, Sira M. New capillary bed formation with a surgically constructed arteriovenous fistula. Plast Reconstr Surg 1980;66(1):109.

[204] Tanaka Y, et al. Generation of an autologous tissue (matrix) flap by combining an arteriovenous shunt loop with artificial skin in rats: preliminary report. Br J Plast Surg 2000;53(1):51−7.

[205] Mian R, et al. Formation of new tissue from an arteriovenous loop in the absence of added extracellular matrix. Tissue Eng 2000;6(6):595−603.

[206] Lokmic Z, et al. An arteriovenous loop in a protected space generates a permanent, highly vascular, tissue-engineered construct. FASEB J 2007;21(2):511−22.

[207] Novosel EC, Kleinhans C, Kluger PJ. Vascularization is the key challenge in tissue engineering. Advanced drug delivery reviews 2011;63(4):300−11.

Fabrication of Artificial Bacteria for Targeted Drug Delivery

11

U Kei Cheang,[1] Min Jun Kim[2]

[1]*Drexel University, Department of Mechanical Engineering & Mechanics, Philadelphia, PA, 19104, USA;* [2]*School of Biomedical Engineering, Science & Health Systems, Drexel University, Philadelphia, PA 19104, USA*

CONTENTS

Biofabrication.

INTRODUCTION

Developments in microrobotics have facilitated research into using miniaturized devices for biomedical reasons. In particular, there is much interest in applying microscale locomotive principles to drug delivery and therapeutic mechanisms [1−6]. The goal is to develop methods and technologies to maximize therapeutic value for targeted delivery and minimizing potential side effects. The development of cancer treatments from traditional chemotherapy injection to target delivery highlights the advantages and the need for high-precision drug delivery. In essence, a microscale robot with the ability to navigate inside the human body can swim through the bloodstream, penetrate mucosal and epithelial barriers, and move through soft tissues to reach the targeted destination.

Challenges in microscale locomotion have created many unanswered questions due to the difficulties in microscale control and manipulation. Alas, there are no accepted standards for microlocomotion given the restrictions in microscale. Nature-inspired principles have proven effective to overcome these limitations. In particular, the propulsion system of flagellated bacteria has been popular in the field of bioinspired microswimming. Much work has been done to study microlocomotion in Newtonian media, and recently researchers are diving into more realistic scenarios by studying swimming in complex environments that are similar to biological fluids such as blood or mucus.

Living tissues are complex media composed of biopolymers, deformable cellular structures, and viscous fluid. Unlike conventional Newtonian media, tissues are non-Newtonian, meaning they exhibit viscoelasticity. The differences between the two types of media can be examined in their reaction to stress in flow: non-Newtonian media have a nonlinear stress-strain relationship and Newtonian media have a linear behavior. As a result, the mechanisms to swim in complex media should be adaptable to the nonlinearity. Various attempts to fabricate and control microscale robots have yet to demonstrate swimming in complex environments. As the idea behind such navigation is potentially feasible through adaptation to the surroundings, this chapter explores the concept of a hybrid biotic/abiotic microswimmer device that utilizes bacterial flagella as a means for propulsion and adaptation. As observed from bacteria, their flagella undergo polymorphic transformation in order to change shape accordingly to survive and swim in different environments, including biofluidic environments. Consequently, bacteria navigate in low Reynolds number media under a wide range of chemical (pH, salinity, etc.) and flow conditions.

11.1 Drug delivery

In recent years, scientists, clinicians, and engineers have been working to overcome the challenges set forth by the limitations in medical and biomedical technologies.

Advances in nanotechnology and bioengineering hold promise to overcome these challenges and revolutionize the medical industry. Of particular interest is drug delivery. Researchers believe that the ability to accurately and precisely deliver needed medicine to a targeted area with a monitored dosage can significantly reduce side effects, complications, and ineffectiveness from conventional delivery methods such as injection. For instance, the active pharmaceutical ingredient (API), or simply the administrated drug, ideally accumulates in the targeted area. However, the drug, enclosed in an excipient, is faced with various obstructions in the body. This results in drug accumulation in the tissues and organs en route to the targeted region, causing negative side effects to the body [7].

11.1.1 Active and passive drug delivery

The concept behind targeted drug delivery is recognition of the targeted area. Current targeted drug delivery techniques are separated into two categories: passive drug delivery and active drug delivery. In passive drug delivery, drugs are usually introduced into an infected area while having a high affinity to the targeted tissues and organs. The drug is accumulated to the target through enhanced permeability and retention (EPR). EPR is the preferential accumulation of molecules such as liposomes, macromolecular, or nanoparticles on target tissues [8]. This delivery strategy works very well for tumors, infarcts, and inflammation areas [7]. Through EPR, the mechanism for drug release depends on the surrounding environment. Different drug carriers, for instance, can be designed to be sensitive to specific ranges of pH or temperatures, providing control over drug accumulation in the target. In order to prolong circulation time for sufficient drug accumulation and control drug concentration, the surface of the drug carrier is usually chemically modified [9,10]. However, specificity is lacking in cases where the targeted sites have similar environmental conditions as normal tissues.

Active targeted delivery enhances the targeting capability through the use of targeting moieties. A number of approaches have been reported. Immunotoxins were used to target the appropriate antigen from cancer cells, while antigen-free cells were free from detection [11]. Enzyme−antibody conjugates were studied for treating thrombolysis [12]. In addition, many body components can be targeting using this approach, including cardiovascular systems, reticuloendothelial systems, lymphatic systems, tumors, infarcts, inflammations, infections, and transplants [7]. In short, the use of target moieties has proven to be relatively effective, though direct administration is still difficult.

11.1.2 Targeted drug delivery using microrobotics

Existing drug delivery methods provide the means for drug administration on targeted tissues and organs. However, due to the nature of the navigation, targeting, and drug releasing mechanism, current methods have limitations in direct administration on the targeted tissue. Recent advances in the field of microrobotics hold much promise in combining microlocomotive principles and control theories to enhance the navigation capability of drug carriers. In principle, miniaturized robots can be controlled to navigate inside a human body and deliver a controlled dosage directly to the targeted tissue or organ. Currently, much work has been done to study microswimming in low Reynolds number conditions, which serves as a precursor to future research to utilize such approaches for in vitro applications and, eventually, in vivo applications.

To develop a microrobot that can perform drug delivery, a number of criteria must be met. The microrobot must be powered by a wireless source, able to swim in complex biofluidic environments, and able to penetrate soft tissues. Fabrication of these microrobots is limited by the micro- and nanofabrication technologies available today; however, a number of research groups have demonstrated successful fabrication using nonconventional methods. A number of wireless power methods have been explored, but considering the potential complications of using chemical and electrical stimulation in biomedical applications, magnetic control holds great promise due to fast response times and minimal health impacts. Recently, more systematic attempts have been made to study navigation in non-Newtonian fluids and microrobot interaction with cells and tissues; however, the successes of such studies are preliminary.

The development of active-controlled drug delivery mechanisms for microrobotics has been largely based on either chemical or mechanical principles. The propulsion mechanisms of interest are based on our understanding of microbial swimming in Newtonian fluids. For drug delivery, however, microrobots have the ability to swim in the bloodstream and penetrate complex materials like mucus, epithelium, and tissues. These challenges remain as roadblocks in applying targeted drug delivery, gene therapy, ionizing radiation, or hyperthermia [13–22]. Microorganisms have continuously demonstrated the ability to achieve effective propulsion through a variety of media during microbial infection. *Helicobacter pylori* can swim through gastric mucus in the interior of the stomach [23]. *Treponema pallidum* and *Borellia burgdorferi* can penetrate through tissue and give rise to syphilis and Lyme disease infection, respectively [24,25]. Therefore, bioinspired propulsion mechanisms are prime candidates for the development of microrobotic drug delivery systems.

While the importance of propulsion in complex media has been established, few studies provide in-depth analyses. There are several theoretical studies, but they lack biological relevance to realistic scenarios. Infinite swimmers undergoing wavelike swimming strokes have been studied [26–29], but finite-length swimmer models provide more realistic analyses [30,31]. Experiments have established the fact that propulsion in complex media is qualitatively different due to changes in swimming speeds [24,32–35], gaits [36], or even swimming strategies [23]. However, well-controlled experiments are needed to validate and improve our understanding of the hydrodynamics of locomotion in such media.

A number of research groups are actively developing microscale robots. A sperm-like microswimmer [37] and an artificial nanocoil swimmer [38] have been experimentally reported to achieve controlled propulsion by using oscillating transverse magnetic fields. Other microrobotic swimmers have also been fabricated and controlled through the use of magnetic field gradients [39–41]; however, the use of gradient mandates that the field must be strong in order to effectively direct motion for magnetic nanoparticles [42,43]. There are also studies on chemical-driven microswimmers [44–46], but the need for a specific engineered chemical environment is not ideal for drug delivery or other biomedical applications due to potential side effects. Nevertheless, transporting nanoparticles through soft tissues remains inefficient and poorly understood [47–52]. However, all of the microrobotic swimmers did not demonstrate swimming in complex fluid or penetration of tissue-like material.

The microrobot described in this chapter attempts to closely mimic the actual bacteria and their flagellar properties. By using the flagella as one of the components and the actuator for propulsion, these microswimmers can directly utilize flagellar hydrodynamics and polymorphic properties employed by bacteria to swim [53]. The microswimmer is actuated and controlled through a rotating magnetic field [54,55]. In essence, this microswimmer will inherit bacteria's ability to navigate in low Reynolds number and complex media and to penetrate soft tissue. It has been reported that magnetic nanoparticles can be given a sufficient torque for rotation by using a small field (1–10 mT) [56,57], which means that patients undergoing treatment would experience less intense magnetic fields, lowering the possibility of side effects or complications from intense magnetic field exposure. Also, the microswimmers will be able to propel themselves [58] through tissues, minimizing the accumulation of particles near the desired target [42].

11.1.3 Interaction with biomechanical microstructures of soft biological tissues

Navigation in a human body is a great challenge due to the environment that the microrobot would encounter. As mentioned previously, many microorganisms have the ability to overcome harsh conditions; therefore, understanding the interaction between microorganisms, soft biological tissues, and other biomedia is crucial in developing artificial microrobots for drug delivery.

Many microorganisms have the ability to swim through complex biomaterials. Sperm can travel through mucus in the female reproductive tract and bacteria can penetrate mucous layers in the respiratory and digestive tracts during infection and disease. Due to the small scale of microorganisms, the microstructural features of biomedia can be significant influences, since the scale of such features is similar to that of microbes. At the microscale, the mechanical deformation, long- and short-range hydrodynamic forces, and contact forces (including van der Waals, electrostatic, and chemical binding) are factors that must considered; at the microscale, transportation through biomaterials depends on swimming speeds and adhesion lifetimes influenced by all these microscale interactions. Spirochetes in gelatin have exhibited different modes of locomotion depending on the adhesion of the medium. A theoretical approach to understanding the effect of microstructural features [59] is by modeling a dynamic phase interacting with both the solvent phase and the swimmer. As a result of this work, it was found that the interaction between swimmers and the microstructure can significantly affect swimming properties [59,60] due to the damping force exerted on a swimmer from mechanical, chemical, electrostatic, and van der Waals interactions between the swimmer and the medium structures.

11.2 Bioinspired locomotion and microswimming

Using microrobots for targeted drug delivery systems must be designed to overcome low Reynolds fluidic environments, complex fluidic environments, soft tissue obstacles, and adverse flows. In order to swim in a low Reynolds number, the microrobot must comply with microfluidic transport phenomena, which is different from the macroscale transport [61]. Therefore, many of the conventional macroscale swimming strategies will be ineffective [58]. Fortunately, several studies have been done on swimming mechanisms that are effective in low Reynolds numbers [58]. However, much research must be done to fully understand the mechanism of navigating through complex media and soft tissue.

Bioinspired propulsive systems hold great promise, mainly due to the abilities demonstrated by microorganisms. As more about microorganisms is learned, there are great potentials to mimic their ways of swimming and apply them to practical engineering systems. Bacterial propulsion has been a prominent choice due to the simplicity and the effectiveness of their propulsion. It is also relatively easy to mimic and model. The concept behind the microrobot discussed in this chapter is to directly utilize the bacterial flagella as a component for the microrobot to harness the properties of the flagella to swim in various environments.

As mentioned, inspiration was drawn from flagellated bacteria [62] by utilizing the polymorphic transformation of bacterial flagellar filaments [53] to adapt to the fluidic environment and for direct fluid actuation at low Reynolds numbers. Bacterial flagella also exhibited excellent mechanical properties, such as stability and durability, making them an ideal material for use in artificial microrobots. Further, their ability to self-assemble from monomers of flagellin protein into long filaments, their suitability for chemical modification and functionalization, and their potential for genetic engineering of their structure [55,63] allow them to be used for engineering systems. More important, bacterial flagella undergo polymorphic transformation both in loaded and unloaded conditions due to chemical, electrical, thermal, mechanical, or optical influence; this means the flagella filament can change into a preferential shape under different conditions to swim adaptively.

11.2.1 Flagella

Bacteria such as *Salmonella typhimurium* or *Escherichia coli* can swim in low Reynolds number environments by rotating their long, thin helical flagella via their flagellar motors [64]. Bacterial flagella are made up of subunits of monomer protein assembled into spiral nanostructures [65]. A flagellar filament is 20 nm in diameter and approximately 10 μm in length. Their primarily function is propulsion [62,66]. Flagella are extremely stiff, having an elastic modulus estimated to be 10^{10} N/m^2. Along with their polymorphic properties, flagella are suited to be used as bionanomaterial. In this chapter, flagella are harnessed as actuators for robotic microswimmers.

Due to their makeup, flagellar filaments are capable of taking numerous polymorphic forms. When exposed to certain changes in the environment, such as pH, flagella respond by undergoing polymorphic transformations, changing their helical handedness and pitch. For example, lowering the pH from 7 to 5 or raising it from 7 to 9 will cause filaments to change from a normal to a coiled form [53]. During the transformation, the molecular monomers, called flagellin, that form longitudinal rows along the length of the entire filament realign in the order of subnanometer. This generates changes in strain that result in changes in filament length by factors as large as 3. Bacterial flagella can also undergo polymorphic transformation both in loaded and unloaded conditions when introduced to chemical, electrical, thermal, mechanical, or optical stimulation. These dynamic properties of filaments are the characteristics that allow bacteria to swim adaptively. In addition, flagellar filaments have remarkable durability and stability, having the capacity to withstand high temperatures (60°C) and extreme pHs (7 ± 4) [67−69]. This allows them to adapt to a wide range of extreme environmental conditions.

In addition to their polymorphic transformation capabilities, the ability to self-assemble in vitro and their surface properties make them fully utilizable as material for nanoactuators for microrobotic swimmers. Flagella are naturally around 10 μm and can be mechanically sheared off. Once sheared, the flagella are generally "broken" into filaments of approximately 5 μm. Despite the shortened length, the flagella's self-assembling ability allows for a certain degree of control over the lengthening of the flagella through polymerization or flagellin monomers. By adjusting the environmental conditions, such as salinity, ionic strength, NaCl content, temperature, and pH, the length and morphology of the flagella can vary greatly [69−70]. Evidently, reconstituted filaments can be polymerized up to 70 μm [53].

The flagellar filament is a tubular structure that consists of 11 strands of protofilaments, each of which is a polymer thread of flagellin [65]. These protofilaments have two distinct subunit conformations, and the ratio of the conformational states in a filament determines the polymorphic helical conformation—that is, the polymorphic shape of the flagella. When the protofilaments of the same type are adjacent to one another, the mechanical strain is minimized, leading to the prediction of 12 polymorphic forms [71,72]. These polymorphic conformations can be triggered by chemical, mechanical, and thermal stimuli [68,69,73,74] and are completely reversible by simply reversing the applied environment stimuli. This shape-changing ability demonstrates great potential for using flagella for engineered applications. Flagella have also proven to be extremely easy to obtain and are cost efficient.

The surface property of the flagella allows for surface functionalization, which is another important aspect that can be taken advantage of for different engineering applications [75]. This is possible due to the peptide compound on the surface of the flagella, providing the basis for surface functionalization of the flagella. This chapter later discusses the biotinylation of flagella through the use of NHS-esters. This process is also known as primary amine biotinylation. In essence, functionalized flagella are modified for particle attachment, which in turn can be used for engineering applications—in this case, for the fabrication of a microrobotic swimmer [53,54].

11.2.1.1 *Spiral waveform*
In the microfluidic condition, locomotion greatly differs from macroscale due to the different perception of forces and mechanics. In microscale, viscous forces and adhesive interactions dominate gravitational and inertial forces due to the dramatic difference in length scales [38,58]. The dynamics of this force relationship can be expressed in a dimensionless characteristic quantity called a Reynolds number. This number is defined as the ratio of inertial forces to viscous forces. In a fluidic condition, the Reynolds number takes into account the viscosity and density of the fluid medium and the length of the scale of interest. The viscosity and density are intrinsic properties of the fluid and are not subject to change unless the medium is changed. The length scale depends on the size of the swimming entity, which is a direct influence on the inertia force due to the mass of the object. Microscale locomotion falls into the low Reynolds number regime. As a result of the highly viscous condition, conventional macroscale swimming strategies such as paddling are ineffective. Fortunately, nature has figured out ways to navigate in microscale, which is demonstrated in microorganisms that use the viscous surroundings for propulsion.

Single-flagellated bacteria such as *Caulobacter crescentus* and *Rhodobacter sphaeroides* propel themselves by simply rotating their flagella. Through rotation, their flagella conform into a spiral waveform and achieve propulsion. As mentioned previously, the shape and form of the flagella are determined by the structural makeup of the flagellin protein. Therefore, the handedness and helical parameters are already predetermined by the various environmental factors such as pH. Multiflagellated bacteria such as *Escherichia coli* and *Salmonella typhimurium* follow the same principle for swimming, but their multiple flagella are used all together as a single effort to propel the cell. As the flagella rotate counterclockwise, they bundle together and push the cell forward. Likewise, the principle is the same as the single-flagellated cells—that is, the spiral waveform used for propulsion. Through this waveform, bacteria employ corkscrew swimming, or spiral swimming [58,76], to travel in low Reynolds numbers as well as complex media environments. As foreseen by Lauga and Powers, the understanding of the evolutionary process of swimming at low

Reynolds environments is vital to designing and optimizing artificial swimmers [61]. Hence, mimicking the swimming strategies of microorganisms is a good approach for designing artificial microrobotic swimmers for drug delivery systems.

11.2.1.2 Bioenvironment navigation

Given the size of microscale swimmers and their perspectives, the mechanical details of microstructures in the medium need to be considered carefully. Unlike homogeneous solutions of polymers, such as methylcellulose, many biomaterials are spatially heterogeneous in the microscale. For instance, when sperm cells swim through viscoelastic mucus in the female reproductive tract, they encounter a homogeneous environment consisting of voids of microstructural networks of the medium [60]. As it turns out, the size of the mucin network mesh near ovulation (up to 25 μm [60,77]) is suitable for sperm to reach the ovum [78]. Therefore, it is clear that homogeneous continuum models that ignore the microstructure of the mucin fibers are likely to be inaccurate and unrealistic. Depending on the menstrual cycle, the mesh size can be large or small (1 μm or smaller). Long-range hydrodynamic interaction between the swimmer and the network is important when sperm swim through a sparse network with a large mesh size and move through the voids, whereas short-range hydrodynamic interaction, such as adhesion or electrostatic forces, is important when swimming through a small mesh. Through this example, it can be seen that the mechanics of the microstructural feature directly influence the forces involved, and the mechanical response of complex biological media should be investigated.

To characterize such biomedia from the microscale, first, the compliance should be investigated based on the deformation in response to stress generated by swimming bacteria. Recent studies on nematodes in granular media have shown the effect of swimming due to this kind of deformation when the granules are pushed aside, leaving a voided path [34,35]. Second, the mechanics of the microstructural components in the medium that generate the macroscale material properties must be well understood. Third, the adhesion forces have been observed to influence spirochete locomotion through tissues and gelatin models [79,80].

11.3 Materials and methods

This section discusses the fabrication, control, and analysis of the microrobotic swimmer that is fabricated using flagellar filaments sheared off from bacteria. These microrobots are designed to mimic bacteria by harnessing their flagella; hence, they are considered artificial bacteria and are called "microswimmers" [54]. Due to their size, conventional microfabrication methods are not used for the fabrication, but rather, the microswimmers were created using nonconventional fabrication methodologies through chemical conjugation. The microswimmers are magnetically actuated and have demonstrated propulsion in a fluid medium. Diffusion was a significant influence on the mobility of the microswimmers; therefore, the analysis of the microswimmers' mobility was done through calculation of the mean squared displacement (MSD).

11.3.1 Microswimmer fabrication

Bacteria are ideal models for the design of abiotic microswimmers. Their structure and propulsion system are relatively uncomplicated and easy to model when compared with other eukaryotic cells. Recent work at Drexel University [54,55] has demonstrated the key processes necessary to fabricate these artificial bacteria. A microswimmer consists of beads with diameters of 200 nm to 3 μm linked by a flagellar filament using avidin-biotin linkages. First, polymerized flagella were biotinylated at the amino groups on the surface of the filaments using NHS (*N*-hydroxysuccinimide) linked to biotin. The biotinylated filaments were depolymerized into biotinylated monomers and then repolymerized into biotinylated seed particles (<1 μm filaments), with sequential addition of nonfunctionalized monomers to create filaments with biotin functional groups at one end. Additional biotinylated monomers were then added to polymerize the second functionalized end. This process creates filaments with biotin functionalized at both ends. Next, polystyrene (PS) beads and magnetic nanoparticles (MNPs) were surface functionalized using purified avidin protein. Finally, the biotinylated flagellar filaments and the avidinated beads (PS beads and MNPs) are conjugated to create microswimmers.

11.3.1.1 Flagella isolation and purification

The flagellar isolation and purification procedure was inspired by the one used by Hesse [63]. Filaments are removed from bacterial cell bodies and reconstituted in a stepwise process. Bacteria were grown in Lysogeny Broth (LB). Their flagellar filaments were harvested by mechanical shearing and centrifugation and then depolymerized into a monomer of protein called flagellin. When filaments are detached from cells, they are generally "broken up" into smaller filaments; they are generally shorter than 10 μm, and the distribution of lengths is wide. When the filaments are repolymerized, long flagella filaments can be reconstituted in vitro in the range of 10 to 25 μm, with some as long as 70 μm. Afterward, they can be stored for months in the polymerization buffer at 4°C.

The bacteria *S. typhimurium* were cultured using 250 mL of LB inoculated with 1 mL of frozen *S. typhimurium* stock. The culture was then incubated at 33°C for 16 hours, with constant shaking to maximize aeration and allow saturation. The saturated culture was then pelleted by centrifugation for 35 minutes at 4500 *g* and then resuspended in 1 mL of polymerization buffer (0.01 M potassium phosphate buffer [pH 6.5], 150 mM NaCl). The resuspended bacteria were then deflagellated by vortexing at varying positions and orientations for 12 minutes. The deflagellated cells bodies were pelleted out by centrifugation at 10,000 *g* for 15 minutes. The flagella in supernatant were taken out and place in an ultracentrifuge for 1 hour at 100,000 *g* and 4°C to remove small debris. The resulting pellet was resuspended in

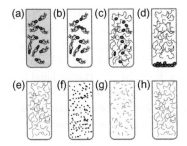

FIGURE 11.1

The flagellar filament isolation process. (a) Bacteria are cultured in LB. (b) The culture is resuspended in a polymerization buffer. (c) The resuspended cells are vortexed to shear off their flagella. (d) The suspension is centrifuged to pellet the cell bodies. (e) The supernatant is retained for the purified filaments. (f) The filaments are depolymerized into monomers. (g) Monomers are used to create flagellar seeding particles. (h) Full-length filaments are polymerized from seeds and monomers.

FIGURE 11.2

Flagellar filament visualized using SEM.

0.5 mL of polymerization buffer and centrifuged at 10,000 g for 15 minutes for further purification of the remaining cell bodies. The resulting supernatant was centrifuged at 100,000 g once again and then resuspended in 0.5 mL of polymerization buffer. The suspension contains flagellar filaments that are 1 to 5 μm long. Figure 11.1(a–h) shows a detailed step-by-step schematic of the flagella purification procedure. The image of a flagellar filament in Figure 11.2 was taken using a scanning electron microscope (SEM).

11.3.1.2 Avidin-biotin chemistry

Avidin is a tetrameric protein found in egg whites. It contains four identical subunits that serve as a binding site for the biotin group, or vitamin B7. The avidin-biotin bonding is the strongest naturally found noncovalent bond [81]. Biotin is widely used in biochemical assays because of its small size and high protein functionality capabilities [82]. The process of affixing biotin molecules onto a protein, nucleic acid, or other molecule is called biotinylation [83]. Avidin has a great affinity for biotin (dissociation constant, $K_d = 10^{-14}$). Biotinylated proteins, such as biotinylated flagella, can be purified and exploited because of this high-affinity bonding. Furthermore, avidin has a high resistance to denaturation in extreme conditions due to its tetrameric structure; therefore, it is sustainable in harsh conditions similar to inside the human body.

Avidin-biotin chemistry has been reported in the past for biomaterial immobilization. This bond is very stable and can withstand a wide range of chemical pH fluctuations. This is important in maintaining structural integrity during the fabrication procedure. Orth's work on immobilizing DNA protein on a silicon substrate [84] showed the potential to use avidin-biotin chemistry for bioengineering proposals. Consequently, Orth's work became an inspiration to develop a procedure for attaching flagella to micro- and nanoparticles.

11.3.1.3 Functionalization of flagella

Flagellar filaments were biotinylated specifically at the amino groups of the entire surface of the filament using NHS (*N*-hydroxysuccinimide) linked to biotin and then incubated for 3 hours at room temperature with constant shaking. Then, the filaments were depolymerized into biotinylated monomers using a water bath at 65°C. The biotinylated monomers were then mixed with repolymerization buffer (2 M Na_2SO_4 in potassium phosphate buffer, pH 6.5) and incubated at room temperature for 1 hour in a shaker. Nonbiotinylated monomers were added to the biotinylated seeds and incubated for 36 hours in room temperature in a shaker. The flagella will be polymerized unidirectionally [85], yielding flagellar filaments that are biotinylated only at one end of the flagellar filament. Next, additional biotinylated monomers were added and then incubated for 12 hours at room temperature. This reaction will yield flagellar filaments that have biotin groups only at both ends of the filament. The detailed flagellar functionalization procedure is shown in Figure 11.3(a–f).

11.3.1.4 Functionalization of particles

To effectively functionalize a magnetic bead with avidin, the surface of the bead must first be chemically modified. The bead is first plasma treated to remove excess carbon or other molecules from the surface. The bead surface is then treated with a 3-aminopropyltriethoxysilane (APTES), incubated for 10 minutes, and rinsed with isopropanol. This allows an amino group to attach to the surface. Glutaric acid is then introduced onto the APTES-treated beads to create a glutaraldehyde layer [86] that acts as a crosslinker for avidin; it is then incubated for 2 hours. Afterward, the beads are thoroughly rinsed with PBS. Finally, avidin

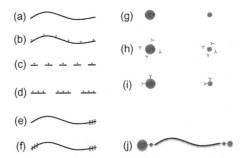

FIGURE 11.3

Microswimmer fabrication. (a) Filaments are isolated from bacteria culture. (b) Flagella filaments are biotinylated. (c) Biotinylated flagella are depolymerized into monomers. (d) Monomers are repolymerized into biotinylated seeding particles. (e) One-ended biotinylated flagella are created by introducing nonbiotinylated monomers. (f) Two-ended biotinylated flagella are created by introducing additional biotinylated monomers. (g) Nonfunctionalized PS beads and MNPs. (h) Avidin is introduced to the particles. (i) The particles are avidinated. (j) Dumbbell microswimmers are fabricated by assembling two-ended biotinylated flagella and avidinated particles.

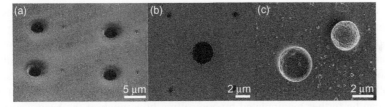

FIGURE 11.4

(a) A 2 × 2 array of nanowells created using the FIB. The larger wells are for the 3 μm polystyrene beads, and the small wells are for the 150 nm magnetic beads. (a) A well pattern for fabrication of multiflagellated microswimmers. (c) SEM image of a polystyrene beads settled inside a well (left) compared with a polystyrene bead not in a well (right).

solution is introduced onto the surface of the chemically modified beads and then incubated for 24 hours in a humid environment to prevent evaporation.

One of the concerns with microswimmer fabrication is the location of flagellar attachments and the orientation of the MNP, which determines the configuration of the microswimmers. Simply mixing the functionalized bead and flagella together does not permit control over the conjugation, resulting in an assortment of configurations. To achieve control over the configuration of the microswimmers, a manufacturing process can be used to fabricate microswimmers using self-assembly methodologies. This method involves the fabrication of patterns or nanoarrays on glass chips using the focused ion beam (FIB), as shown in Figure 11.4. Beads can be separated to fall into their perspective wells through size separation, and the excess beads should be washed away from the surface of the glass. Then biotinylated flagella will be introduced and allowed to attach to the beads. Once the fabrication is finished, the microswimmers can be inserted via agitation without incurring damage to their structure. To fabricate these wells, a glass chip is first coated with an approximately 3.8-nm layer of platinum/palladium using a sputter coater. This layer is necessary to focus the ion beam from the FIB/SEM system and drill the array of micro- and nanowells. After drilling, the bottom of the well is glass, while the rest of the surface of the chip is coated with metal. Finally, a monolayer of perfluorodecyltrichlorosilane (PFDTCS) is deposited onto the surface using chemical vapor deposition (CVD). The silane will bind to the glass only and not the metal. This allows for preferential binding of beads inside the wells, while beads on the surface of the glass chip that are on metal can be washed away. Figure 11.4c shows beads settled inside the wells.

11.3.2 Microswimmer structures

The biotinylated flagellar filaments and the avidinated beads are combined in a reaction mix and incubated for 5 hours at room temperature on a shaker. This will yield an assortment of microswimmer configurations. The configuration we discuss here is the dumbbell structure. The SEM image in Figure 11.5a shows a single flagellum connected to a bead, and Figure 11.5b shows multiple flagella connected to each bead. This demonstrates the successful fabrication conjugation of beads to flagella. In the FIB nanowell method, biotinylated flagellar filaments also can be added to the beads that settled inside the well. The flagella will attach only to the exposed side of the beads, and the orientation of the beads can be controlled through an applied magnetic field. The location of flagellar attachments can be further manipulated by controlling how much of the beads' surfaces are exposed by changing the depths of the wells. Two configurations of the

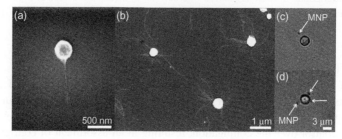

FIGURE 11.5

Flagella attached to beads using the avidin-biotin chemistry. (a) Single flagellum attached to a bead. (b) Multiple flagella attached to each bead. (c) Single-flagellated dumbbell microswimmer. (d) Multiflagellated microswimmer.

microswimmer are shown in Figures 11.5c and d: dumbbell microswimmers, and multiflagellated microswimmers.

11.3.2.1 Dumbbell helical swimming

The dumbbell microswimmer consists of a PS bead connected to an MNP via a flagellar filament. Once the MNP rotates, the flagellar filament connected will rotate as a helix, and its shape will be determined by the polymorphic form of the filament under the specific environment conditions. When sufficient torque is applied, the entire microswimmer will rotate. This will create propulsion similar to the propulsive mechanism of bacteria. However, one must consider the counterrotation of the cell body for a live bacterium. In the case of the microswimmer, the beads rotate in the same direction as the flagella. Despite this, the flagella hydrodynamics should closely mimic that of live bacteria.

The dumbbell swimmers were observed under an enclosed environment in a PDMS fluidic cell with no flow. A Helmholtz coil system is employed as a wireless energy source by generating a rotating magnetic field via two pairs of electromagnetic coils. A 3D representation and the profile of the magnetic field are shown in Figure 11.6.

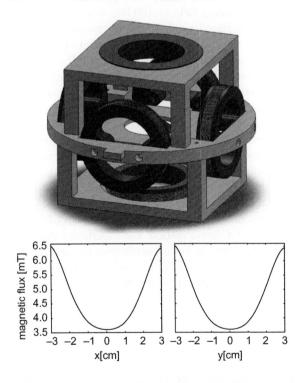

FIGURE 11.6

Top image shows a 3D drawing of the Helmholtz coil system. The plots on the bottom show the profile of the magnetic field, demonstrating homogeneity at the very center.

The flagella act as actuators that convert rotational motion of the swimmer to translational motion of the artificial bacteria. Upon actuation, the rotation of the flagellum will generate a spiral waveform needed to generate the hydrodynamics for swimming.

The control system consists of the Helmholtz coil that is mounted on a Leica DMIRB inverted microscope. The coils are connected to three power supplies for 3D control. A DAQ controller and LabVIEW interface are set up to generate sinusoidal inputs to two of the power supplies with a 90-degree phase lag between them. These inputs allow the coils to generate a homogeneous rotating magnetic field at the center. Under the rotating field, the microswimmers will rotate and swim.

The microswimmers examined were created using 1 μm PS beads and 300 nm MNPs. As a benchmark, the microswimmers were first observed in the PDMS fluidic chamber without applying a rotating field. As expected, the microswimmer demonstrated Brownian motion due to the diffusion of the PS microbead, the MNP, and the flagella. The diffusion of the microswimmer also accounts for the internal dynamics of the change in distance between the PS microbead and the MNP; as a result, the microswimmers have enhanced diffusion because of the bead-to-bead hydrodynamic interaction [87] when compared with a single bead.

In this study, a rotating field of 2.2 mT was applied in the XZ planes with a rotational frequency of the rotating field of 100 Hz, which corresponds to the typical frequency of the bacterial flagellar motor [88]. In this case, the microswimmer will be expected to swim in the Y direction. The movement of the microswimmers was readily apparent when compared to the benchmark test. The microswimmers moved in a positive Y direction with an average speed of 1.3 μm/s. The response of the microswimmers was consistent unless the microswimmers had sediment and formed some type of attachment to the substrate. The diffusion of the microswimmers in a no-field condition (benchmark) was compared with the diffusion of the swimmers under the rotating field. An example of the path of a microswimmer is shown in Figure 11.6. After recording and tracking numerous microswimmers, their ensemble displacement was computed and used to analyze their motility. The plot in Figure 11.7 shows the MSD of the benchmark compared with the actuated microswimmers. Looking on the correlation at a small time scale (t <1 s), the microswimmers exhibited ballistic motion. At a long time scale (t >1 s), ballistic-driven motion changed into translational diffusive motion. The experiment shows that the diffusion of the actuated microswimmer is much greater than the benchmark. This signifies superdiffusion, and the increases in diffusion and the directional bias in the desired direction were speculated to be caused by flagellar hydrodynamics.

CONCLUSION

Existing drug delivery methods offer a means to deliver the pharmaceutical agent to the targeted area. However, drug accumulation may occur at undesired

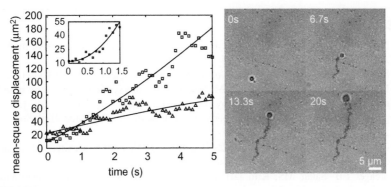

FIGURE 11.7

The plot shows the MSD of actuated microswimmers compared with the benchmark experiment. The four images of the microswimmer show an example of the path taken by an actuated microswimmer over a period of 20 seconds.

locations, which can lead to side effects and complications. Direct drug administration to target tissues or organs proved to be a challenge due to the inaccuracy of current methods. As the field of microrobotics advances, it seems possible to integrate microrobotic research with drug delivery applications. Though the development of a working microrobotic system for drug delivery still has a long way to go, much preliminary work has been done to demonstrate the ability to wirelessly control a microswimming robot to perform simple tasks such as navigation and manipulation. Due to the deeper understanding of the effectiveness of microorganism swimming strategies, many have modeled the bacteria as an approach to achieve drug delivery or other biomedical applications.

It has been established that biomimetic approaches might provide the means to overcome the challenges in navigating in the biological fluidic environment. The artificial bacteria, or microswimmers, discussed in this chapter provide a way of using bacterial flagella as a component to closely mimic not only the swimming mechanism but also the adaptive nature of the polymorphic properties of flagella. After preliminary experiments with these microswimmers, it was observed that the actuation of the flagella created propulsion in the intended direction. However, a much more in-depth examination of swimming in complex media is required to move to the next step toward drug delivery.

Acknowledgment

This work is funded by NSF CBET: Fluid Dynamics Grant.

References

[1] Nelson BJ, Kaliakatsos IK, Abbott JJ. Microrobots for minimally invasive medicine. Annu Rev Biomed Eng 2010;12:55—85.

[2] Sitti M. Miniature devices: voyage of the microrobots. Nature 2009;458:1121—2.

[3] Martel S, Felfoul O, Mohammadi M, Mathieu JB. Interventional procedure based on nanorobots propelled and steered by flagellated magnetotactic bacteria for direct targeting of tumors in the human body. Conf Proc IEEE Eng Med Biol Soc 2008; 2497—500.

[4] Cavalcanti A, Shirinzadeh B, Freitas RA, Hogg T. Nanorobot architecture for medical target identification. Nanotechnology 2008;19:015103.

[5] Steager EB, Sakar MS, Kim DH, Kumar VJ, Pappas GJ, Kim MJ. Electrokinetic and optical control of bacterial microrobots. J Micromech Microeng 2011;21(3): 035001.

[6] Kagan D, Laocharoensuk R, Zimmerman M, Clawson C, Balasubramanian S, Kang D, et al. Rapid delivery of drug carriers propelled and navigated by catalytic nanoshuttles. Small 2010;6(23):2741—7.

[7] Torchilin VP. In: Schäfer-Korting M, editor. Passive and active drug targeting: drug delivery to tumors as an example. Berlin; Heidelberg: Springer; 2010. p. 3—53.

[8] Maeda H, Wu J, Sawa T, Matsumura Y, Hori K. Tumor vascular permeability and the EPR effect in macromolecular therapeutics: a review. J Control Release 2000;65 (1—2):271—84.

[9] Klibanov AL, Maruyama K, Torchilin VP, Huang L. Amphipathic polyethyleneglycols effectively prolong the circulation time of liposomes. FEBS Lett 1990;268(1): 235—7.

[10] Trubetskoy VS, Torchilin VP. Use of polyoxyethylene-lipid conjugates as long-circulating carriers for delivery of therapeutic and diagnostic agents. Adv Drug Deliv Rev 1995;16(2—3):311—20.

[11] Vitetta E, Krolick K, Miyama-Inaba M, Cushley W, Uhr J. Immunotoxins: a new approach to cancer therapy. Science 1983;219(4585):644—50.

[12] Haber E, Bode C, Matsueda GR, Reed GL, Runge MS. Antibody targeting as a thrombolytic strategy. Ann New York Acad Sci 1992;667(1):365—81.

[13] Polyak B, Fishbein I, Chorny M, Alferiev I, Williams D, Yellen B, et al. High field gradient targeting of magnetic nanoparticle-loaded endothelial cells to the surfaces of steel stents. Proc Natl Acad Sci U S A 2008;105(2):698—703.

[14] Chorny M, Polyak B, Alferiev IS, Walsh K, Friedman G, Levy RJ. Magnetically driven plasmid DNA delivery with biodegradable polymeric nanoparticles. FASEB J 2007;21(10):2510—9.

[15] Hafeli UO, Sweeney SM, Beresford BA, Humm JL, Macklis RM. Effective targeting of magnetic radioactive 90Y-microspheres to tumor cells by an externally applied magnetic field. Preliminary in vitro and in vivo results. Nucl Med Biol 1995;22(2): 147—55.

[16] Johannsen M, Thiesen B, Jordan A, Taymoorian K, Gneveckow U, Waldofner N, et al. Magnetic fluid hyperthermia (MFH) reduces prostate cancer growth in the orthotopic Dunning R3327 rat model. Prostate 2005;64(3):283—92.

[17] Lubbe AS, Alexiou C, Bergemann C. Clinical applications of magnetic drug targeting. J Surg Res 2001;95(2):200—6.

[18] Lubbe AS, Bergemann C, Alexiou C. Targeting tumors with magnetic drugs. Tumor Target Cancer Ther 2002;34:379−88.

[19] Pan S, Gulati R, Mueske CS, Witt TA, Lerman A, Burnett Jr. JC, et al. Gene transfer of a novel vasoactive natriuretic peptide stimulates cGMP and lowers blood pressure in mice. Am J Physiol Heart Circ Physiol 2004;286(6):2213−8.

[20] Pislaru SV, Harbuzariu A, Gulati R, Witt T, Sandhu NP, Simari RD, et al. Magnetically targeted endothelial cell localization in stented vessels. J Am Coll Cardiol 2006;48(9): 1839−45.

[21] Pislaru SV, Harbuzariu A, Agarwal G, Witt T, Gulati R, Sandhu NP, et al. Magnetic forces enable rapid endothelialization of synthetic vascular grafts. Circulation 2006;114:314−8.

[22] Polyak B, Friedman G. Magnetic targeting for site-specific drug delivery: applications and clinical potential. Expert Opin Drug Deliv 2009;6(1):53−70.

[23] Celli JP, Turner BS, Afdhal NH, Keates S, Ghiran I, Kelly CP, et al. *Heli cobacter pylori* moves through mucus by reducing mucin viscoelasticity. Proc Natl Acad Sci U S A 2009;106:14321−6.

[24] Berg HC, Turner L. Movement of microorganisms in viscous environments. Nature 1979;278:349−51.

[25] Kimsey RB, Spielman A. Motility of Lyme disease spirochetes in fluids as viscous as the extracellular matrix. J Infect Dis 1990;162:1205−8.

[26] Fu HC, Wolgemuth CW, Powers TR. Beating pattern of filaments in viscoelastic fluids. Phys Rev E 2008;78:041913.

[27] Fu HC, Wolgemuth CW, Powers TR. Swimming speeds of filaments in nonlinearly viscoelastic fluids. Phys Fluids 2009;21:033102.

[28] Fu HC, Powers TR, Wolgemuth CW. Theory of swimming filaments in viscoelastic media. Phys Rev Lett 2007;99:258101.

[29] Lauga E. Propulsion in a viscoelastic fluid. Phys Fluids 2007;19:083104.

[30] Teran J, Fauci L, Shelley M. Viscoelastic fluid response can increase the speed and efficiency of a free swimmer. Phys Rev Lett 2010;104:038101.

[31] Zhu L, Do-Quang M, Lauga E, Brandt L. Locomotion by tangential deformation in a polymeric fluid. Phys Rev E 2011;83:011901.

[32] Turner L, Ryu W, Berg HC. Real-time imaging of fluorescent flagellar filaments. J Bacteriol 2000;182(10):2793−801.

[33] Shen XN, Arratia PE. Undulatory swimming in viscoelastic fluids. Phys Rev Lett 2011;106:208101.

[34] Jung S. *Caenorhabditis elegans* swimming in a saturated particulate system. Phys Fluids 2010;22:031903.

[35] Juarez G, Lu K, Sznitman J, Arratia PE. Motility of small nematodes in wet granular media. Europhys Lett 2010;92:44002.

[36] Suarez S, Dai X. Hyperactivation enhances mouse sperm capacity for penetrating viscoelastic media. Biol Reprod 1992;46:686−91.

[37] Dreyfus R, Baudry J, Roper ML, Fermigier M, Stone HA. Microscopic artificial swimmers. Nature 2005;437:862−5.

[38] Abbott JJ, Nagy Z, Beyeler F, Nelson BJ. Robotics in the small, part I: microbotics. Rob Autom Mag, IEEE 2007;14(2):92−103.

[39] Forbes ZG, Yellen BB, Halverson DS, Fridman G, Barbee KA, Friedman G. Validation of high gradient magnetic field based drug delivery to magnetizable implants under flow. IEEE Trans Biomed Eng 2008;55(2):643−9.

[40] Yellen BB, Forbes ZG, Halverson DS, Fridman G, Chorny M, Levy R, et al. Targeted drug delivery to magnetic implants for therapeutic applications. J Magn Magn Mater 2005;293(1):647−54.

[41] Ghosh A, Fischer P. Controlled propulsion of artificial magnetic nanostructured propellers. Nano Lett 2009;9(6):2243−5.

[42] Dobson J. Magnetic nanoparticles for drug delivery. Drug Develop Res 2006;67 (1):55−60.

[43] Berry CC, Curtis ASG. Functionalisation of magnetic nanoparticles for application in biomedicine. J Phys D Appl Phys 2003;36:R198−206.

[44] Wu J, Kagan D, Balasubramanian S, Manesh K, Campuzano S, Wang J. Motion-based DNA detection using catalytic nanomotors. Nat Commun 2010;1(4):1−6.

[45] Wang J. Can man-made nanomachines compete with nature biomotors? ACS Nano 2009;3(1):4−9.

[46] Gao W, Sattayasamitsathit S, Manian Manesh K, Weihs D, Wang J. Magnetically-powered flexible metal nanowire motors. J Am Chem Soc 2010;132:14403−5.

[47] Barnes AL, Wassel RA, Mondalek F, Che K, Dormer KJ, Kopke RD. Magnetic characterization of superparamagnetic nanoparticles pulled through model membranes. Bio Magn Res Tech 2007;5(1):1−10.

[48] Kalambur VS, Han B, Hammer BE, Shield TW, Bischof JC. In vitro characterization of movement, heating and visualization of magnetic nanoparticles for biomedical applications. Nanotechnology 2005;16:1221−33.

[49] Kuhn SJ, Finch SK, Hallahan DE, Giorgio TD. Proteolytic surface functionalization enhances in vitro magnetic nanoparticle mobility through extracellular matrix. Nano Lett 2006;6(2):306−12.

[50] Kuhn SJ, Hallahan DE, Giorgio TD. Characterization of superparamagnetic nanoparticle interactions with extracellular matrix in an in vitro system. Annu Biomed Eng 2006;34(1):51−8.

[51] Mondalek FG, Zhang YY, Kropp B, Kopke RD, Ge X, Jackson RL, et al. The permeability of SPION over an artificial three-layer membrane is enhanced by external magnetic field. J Nanobiotech 2006;4(4):1−9.

[52] Rotariu O, Udrea LE, Strachan NJC, Badescu V. The guidance of magnetic colloids in simulated tissues for targeted drug delivery. J Optoelectron Adv Mater 2007;9 (4):942−5.

[53] Darnton NC, Berg HC. Force-extension measurements on bacterial flagella: triggering polymorphic transformations. Biophys J 2007;92(6):2230−6.

[54] Cheang UK, Roy D, Lee JH, Kim MJ. Fabrication and magnetic control of bacteria-inspired robotic microswimmers. Appl Phys Lett 2010;97:213704.

[55] Hesse WR, Casale DM, Milton B, Fink PK, Kim MJ. Biologically inspired drug delivery microrobots. In: The 5th international conference on microtechnologies in medicine and biology. Quebec, Canada; 2009.

[56] Tierno P, Goldstanian R, Pagonabarraga I, Sagues F. Controlled swimming in confined fluids of magnetically actuated colloidal rotors. Phys Rev Lett 2008;101: 218304.

[57] Hu S, Eberhard L, Chen J, Love CJ, Butler JP, Fredberg JJ, et al. Mechanical anisotropy of adherent cells probed by a three dimensional magnetic twisting device. Am J Physiol Cell Physiol 2008;287:1184−91.

[58] Purcell EM. Life at low Reynolds number. Am J Phys 1977;45:3−11.

[59] Fu HC, Shenoy VB, Powers TR. Low-Reynolds-number swimming in gels. EPL (Europhysics Letters) 2010;91(2):24002.

[60] Rutllant J, López-Béjar M, López-Gatius F. Ultrastructural and rheological properties of bovine vaginal fluid and its relation to sperm motility and fertilization: a review. Reprod Domest Anim 2005;40(2):79−86.

[61] Lauga E, Powers TR. The hydrodynamics of swimming microorganisms. Rep Prog Phys 2009;72(9):096601.

[62] Berg HC. *E. coli* in motion. Biol Med Phys Series 2003; Berlin: Springer Verlag.

[63] Hesse WR, Luo L, Zhang G, Mulero R, Cho J, Kim MJ. Mineralization of flagella for nanotube formation. Mater Sci Eng C 2009;29(7):2282−6.

[64] Berg HC, Anderson RA. Bacteria swim by rotating their flagellar filaments. Nature 1973;245(5425):380−2.

[65] Calladine C. Construction of bacterial flagella. Nature 1975;255(5504):121−4.

[66] Manson MD, Tedesco P, Berg HC, Harold FM, Van Der Drift C. A protonmotive force drives bacterial flagella. Proc Natl Acad Sci U S A 1977;74:3060−4.

[67] Asakura S. Polymerization of flagellin and polymorphism of flagella. Advan Biophys 1970;1:99−104.

[68] Kamiya R, Asakura S. Helical transformations of *Salmonella* flagella in vitro. J Mol Biol 1976;106(1):167−86.

[69] Kamiya R, Asakura S. Flagellar transformations at alkaline pH. J Mol Biol 1976;108 (2):513−8.

[70] Hasegawa E, Kamiya R, Asakura S. Thermal transition in helical forms of *Salmonella* flagella. J Mol Biol 1982;160:609−21.

[71] Calladine CR. Change in waveform in bacterial flagella: the role of mechanics at the molecular level. J Mol Biol 1978;118:457−79.

[72] Kamiya R, Asakura S, Wakabayashi K, Namba K. Transition of bacterial flagella from helical to straight forms with different subunit arrangements. J Mol Biol 1979;131:725−42.

[73] Macnab RM, Ornston MK. Normal-to-curly flagellar transitions and their role in bacterial tumbling. Stabilization of an alternative quaternary structure by mechanical force. J Mol Biol 1977;112(1):1−30.

[74] Hotani H. Micro-video study of moving bacterial flagellar filaments: III. Cyclic transformation induced by mechanical force. J Mol Biol 1982;156(4):791−806.

[75] Kim M., Hesse W.R., Mulero R., Luo L., Cho J., inventors; Flagella as a biological material for nanostructured devices, 2010.

[76] Keaveny EE, Maxey MR. Spiral swimming of an artificial micro-swimmer. J Fluid Mech 2008;598:293−319.

[77] Lai SK, Wang Y-Y, Wirtz D, Hanes J. Micro- and macrorheology of mucus. Adv Drug Deliv Rev 2009;61(2):86−100.

[78] Suarez SS, Pacey AA. Sperm transport in the female reproductive tract. Human Reprod Update 2006;12(1):23−37.

[79] Moriarty TJ, Norman MU, Colarusso P, Bankhead T, Kubes P, Chaconas G. Real-time high resolution 3D imaging of the lyme disease spirochete adhering to and escaping from the vasculature of a living host. PLoS Pathog 2008;4(6):e1000090.

[80] Harman MW, Dunham-Ems SM, Caimano MJ, Belperron AA, Bockenstedt LK, Fu HC, et al. The heterogeneous motility of the Lyme disease spirochete in gelatin mimics dissemination through tissue. Proc Natl Acad Sci U S A 2012.

[81] Diamandis EP, Christopoulos TK. The biotin-(strept)avidin system: principles and applications in biotechnology. Clin Chem 1991;37(5):625–36.

[82] Chaiet L, Wolf FJ. The properties of streptavidin, a biotin-binding protein produced by streptomycetes. Arch Biochem Biophys 1964;106(0):1–5.

[83] Bayer EA, Wilchek M. Protein biotinylation. In: Meir W, Edward AB, editors. Methods in enzymology. Academic Press; 1990. p. 138–60.

[84] Orth RN, Clark TG, Craighead HG. Avidin-biotin micropatterning methods for biosensor applications. Biomed Microdevices 2003;5(1):29–34.

[85] Asakura S, Eguchi G, Iino T. Unidirectional growth of *Salmonella* flagella in vitro. J Mol Biol 1968;35(1):227–36.

[86] Guesdon JL, Ternynck T, Avrameas S. The use of avidin-biotin interaction in immunoenzymatic techniques. J Histochem Cytochem 1979;27(8):1131–9.

[87] Bammert J, Schreiber S, Zimmermann W. Dumbbell diffusion in a spatially periodic potential. Phys Rev E 2008;77(4):042102.

[88] Darnton N, Turner L, Breuer K, Berg HC. Moving fluid with bacterial carpets. Biophys J 2004;86(3):1863–70.

Biofabricating the Bio-Device Interface Using Biological Materials and Mechanisms

12

Yi Cheng,[1,2] Yi Liu,[3] Benjamin D. Liba,[3] Reza Ghodssi,[2,4] Gary W. Rubloff,[1,2] William E. Bentley,[3,5] Gregory F. Payne[3,5]

[1]*Department of Materials Science and Engineering, University of Maryland, College Park, MD 20742, USA;* [2]*Institute for Systems Research, University of Maryland, College Park, MD 20742, USA;* [3]*Institute for Bioscience and Biotechnology Research, University of Maryland, College Park, Maryland 20742, USA;* [4]*Department of Electrical and Computer Engineering, University of Maryland, College Park, MD 20742, USA;* [5]*Fischell Department of Bioengineering, University of Maryland, College Park, MD 20742, USA*

CONTENTS

INTRODUCTION

In a single generation, electronic devices transformed the way we live our lives. The exponential growth in computing power that catalyzed this transformation was enabled by technological advances in fabricating integrated circuits of ever-greater complexity. Today, a common goal is to expand the power of microelectronics by integrating a broader range of materials into the circuitry and by extending their reach into new applications. In particular, the integration of biological components/systems "on-chip" should enable entirely new capabilities such as multiplexed analysis at the point-of-care, high-throughput screening of drug candidates, biofuel cells to power implantable devices, and the in vitro study of complex biological systems over diverse time and length scales. Yet, there are challenges to integrating biology into electronics: the paradigms for fabrication are entirely different between technological and biological systems, and these paradigms often appear incompatible. Over a decade ago, we set out to create tools for integrating biology with electronics, and our approach was to enlist biology's materials and mechanisms to "biofabricate" the bio-device interface [1−4].

Biology offers several important fabrication capabilities. First, biology is capable of precision manufacturing at the nano-scale. The templated processes of transcription and translation convert genetic information into polypeptide chains of precise sequence and size (i.e., molecular weight). These polypeptides fold into a native three-dimensional conformation that precisely localizes and orients amino acid residues in space. Second, biology self-assembles their nano-scale components (i.e., proteins) over a hierarchy of length scales to generate functional, macroscopic structures (e.g., tissue). Finally, biology employs molecular recognition to allow fabrication to be performed selectively under mild conditions and in complex environments (e.g., in multicomponent aqueous solutions). In some cases, biology enlists molecular recognition to assemble supramolecular structures that are held together through weak physical forces (e.g., assembly of the virus capsid from coat proteins). In other cases, molecular recognition is employed for the enzymatic-introduction of covalent bonds that further stabilize structure (e.g., the crosslinking of collagen fibrils).We envision that these biofabrication capabilities can provide an important complement to traditional methods to microfabricate electronic devices.

12.1 Complementarity of biofabrication and microfabrication

Table 12.1 illustrates that the materials and methods that biology employs for assembly are fundamentally different from those used in the technological processes to fabricate electronic devices. A common feature of biology is the synthesis of modular, nano-scale components (e.g., individual proteins or protein domains) that are then assembled from the bottom up over a hierarchy of length scales. In contrast, microfabrication's top-down methods do not lend themselves to the precision-assembly of such preformed nano-scale components. Potentially, biology can provide important lessons on hierarchical assembly and possibly even

Table 12.1 Divergence or Complementarity Between Biological Fabrication and Microfabrication

	Biological Fabrication	**Microfabrication**
Fabrication paradigm	Bottom-up and hierarchical	Top-down and monolithic
Common materials	Soft (e.g., biopolymers)	Hard (e.g., inorganics)
Approach for controlling chemistry	Discern through molecular recognition	Exclude contaminants through engineering
Approach for controlling defects	Correct and heal defects	Strive for defect-free fabrication
Final structure	Dynamic (for adaptability)	Static (for reliability)

provide the template for assembly [5–7]. Another common feature of biology is the use of soft matter (e.g., proteins and polysaccharides) to create water-rich matrices that are compatible with labile components (e.g., enzymes) and living systems (e.g., populations of cells). In contrast, microfabricated devices are generally composed of hard materials (e.g., silicon and metals) that are less common to biology. The ability to extend the capabilities of microelectronics into the life and medical sciences may rely on the ability to construct a "soft" biocompatible interface with the otherwise hard electronics.

In addition, biology generally creates dynamic structures that can assemble/disassemble in response to endogenous cues or environmental stimuli. Traditionally, microfabrication methods aimed to generate stable (i.e., static) structures to ensure reliable, long-term operation. However, the potential to confer capabilities for error-correction, self-healing, and adaptive (reconfigurable) structures could endow electronic devices with even more powerful capabilities.

12.2 Biofabrication methods

Figure 12.1 illustrates our vision for biofabricating the biology-device interface. We especially focus on (1) triggering the self-assembly of stimuli-responsive hydrogel-forming biological materials, (2) "connecting" functional components through the enzymatic-introduction of covalent bonds, and (3) engineering proteins to promote their assembly.

12.2.1 Self-assembly

Biology is well known for its use of self-assembly to create hierarchical structure that confers function. While there has been exciting research on the use of proteins and nucleic acids for programmable self-assembly, our focus has been on polysaccharides that can be triggered to self-assemble in response to mild

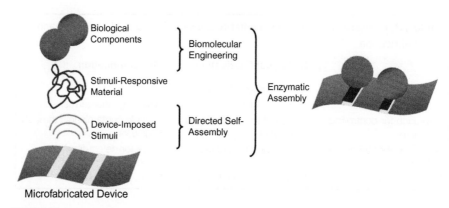

FIGURE 12.1

Vision for enlisting biological materials and mechanisms to build the bio-device interface. Stimuli-responsive materials are triggered to self-assemble in response to device-imposed stimuli while biological components (e.g., proteins) can be engineered to facilitate assembly to the interface.

Source: Adapted from reference [2] with permission from The Institute of Physics Publishing.

environmental stimuli. In particular, we focus on polysaccharides that can be induced to undergo a sol-gel transition in response to local, device-imposed electrical signals. Mechanisms that allow hydrogels to be electrodeposited provide a means to "address" various biological components to electrode surfaces.

To our knowledge, the first biopolymer to be electrodeposited was the pH-responsive film-forming aminopolysaccharide chitosan. Figure 12.2a shows the neutralization mechanism for chitosan's cathodic deposition [8−14]. In this case, electrolysis reactions at the cathode generate a pH gradient, and the high localized pH adjacent to the cathode surface induces chitosan to undergo its reversible sol-gel transition. It appears that the acidic polysaccharide alginate was the next biopolymer to be electrodeposited—in this case by an anodic neutralization mechanism [15]. Alginate is also Ca^{2+}-responsive, and Figure 12.2b shows a mechanism for the reversible electrodeposition of Ca^{2+}-alginate hydrogels. In this case, protons generated by anodic electrolysis reactions serve to solubilize $CaCO_3$ with the release of the Ca^{2+} ions that trigger the formation of the Ca^{2+}-alginate hydrogel network [16,17].

These initial studies demonstrated that the stimuli-responsive gel-forming properties of chitosan and alginate provide a mechanism for the electrodeposition of hydrogel films. Importantly, it has been observed that various components can be added to their deposition solutions and co-deposited with chitosan and alginate [18,19]. Thus, co-deposition provides a simple means to generate films that are functionalized with enzymes (for biosensing [11]), nanoparticles (for signal transduction [20]), and prokaryotic [16] and eukaryotic [21] cells. In addition, co-deposition provides a means to generate composite coatings (e.g., for implants) [12,22−24].

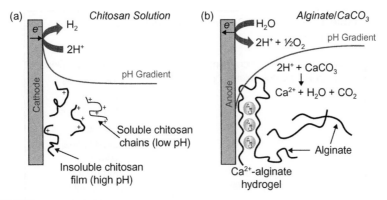

FIGURE 12.2

Mechanisms that allow device-imposed electrical signals to trigger hydrogel self-assembly. (a) Chitosan electrodeposits at a cathode by a neutralization mechanism. (b) Calcium alginate electrodeposits at an anode by the localized solubilization of $CaCO_3$ to generate Ca^{2+} ions.

Co-deposition also provides a means to assemble matrices that respond to orthogonal stimuli. For instance, we used Ca^{2+}-alginate to co-deposit the thermally responsive and neutral polysaccharide agarose [21]. Upon cooling, agarose forms its hydrogel network. Importantly, the alginate-agarose network has a lower-charge density and allows anionic proteins to be electrophoresed through the co-deposited dual network (anionic proteins can be blocked from migrating through an alginate network, presumably because of electrostatic repulsions) [21].

In summary, co-deposition with chitosan and alginate provides a simple, rapid, and reagentless means to enlist device-imposed electrical signals to direct the assembly of biological and nonbiological components to specific electrode addresses. The deposition mechanisms described above are reversible and spatially controllable, and they yield supramolecular polysaccharide networks. While much research aims to apply chitosan and alginate deposition (e.g., for biosensing and coating implants), it is important to note that there are still opportunities to discover new materials (e.g., polylysine and polyornithine [25]) and new mechanisms for electrodeposition (e.g., Fe^{3+}-mediated deposition of alginate [26]), and to couple electrodeposition with other self-assembly methods (e.g., layer-by-layer [27]).

12.2.2 Enzymatic-assembly

Biological-synthesis and microelectronics-fabrication both must control chemistry to minimize undesired side reactions. Biology achieves this feat using enzymes that catalyze reactions with high selectivities, and these enzymes catalyze reactions under mild conditions and without the need for solvents or toxic reagents. Thus, there is considerable appeal to enlisting enzymes for fabrication [28,29]. Despite

Table 12.2 Enzymes Commonly Considered for Building Structure

Enzymes that generate covalent bonds for macromolecular conjugation or crosslinking	Transglutaminases
	Tyrosinases
	Sortase
	Peroxidases
	Laccases
	Lysyl(amine) oxidase
Enzymes that trigger the formation of supramolecular structures	Kinases/ phosphatases
	Proteases
	β-lactamase
	Glucose oxidase

the appeal however, there are few examples of enzymes that can build structure without the need for reactive substrates or complex co-factors (e.g., ATP).

Table 12.2 lists several enzymes that are being examined to create macromolecular structure. We focus on the enzymes listed at the top that introduce covalent bonds to generate macromolecular conjugates [30,31] or networks [32,33]. We also note an emerging area of research in which enzymes (listed at the bottom in Table 12.2) are used to trigger the formation of supramolecular structure (e.g., physical hydrogels) [34]. For instance, an enzyme could be used to cleave a group that confers solubility to the reactant (e.g., a phosphate moiety) to generate a product that undergoes self-assembly.

As illustrated in Figure 12.3, we use two enzymes to build macromolecular structure. Tyrosinases and related phenol oxidases are copper-containing oxidative enzymes that use molecular oxygen to oxidize phenolics—either low molecular weight phenolics or accessible phenolic moieties of macromolecules (e.g., the tyrosine or dihydroxyphenylalanine residues of proteins). These enzymes are ubiquitous in nature and are responsible for familiar activities such as the enzymatic browning of foods, the sealing and hardening of insect cuticles, and the crosslinking of the mussel's adhesive protein. Oxidation converts the phenolics into reactive o-quinones that can leave the enzyme's active site and undergo subsequent uncatalyzed reactions with available nucleophiles. As illustrated by the reaction in Figure 12.3a, we use tyrosinase to convert accessible tyrosine residues of proteins into "activated" quinones that graft onto chitosan's nucleophilic amines. Tyrosinase is active at pHs where chitosan can be either soluble (pH <6) or insoluble (pH >6.5), and thus tyrosinase can mediate protein grafting to soluble chains or chitosan surfaces. When the extent of tyrosinase-mediated grafting is limited, the protein-chitosan conjugate can retain chitosan's pH-responsive film-forming properties. Thus, tyrosinase provides a means to conjugate proteins

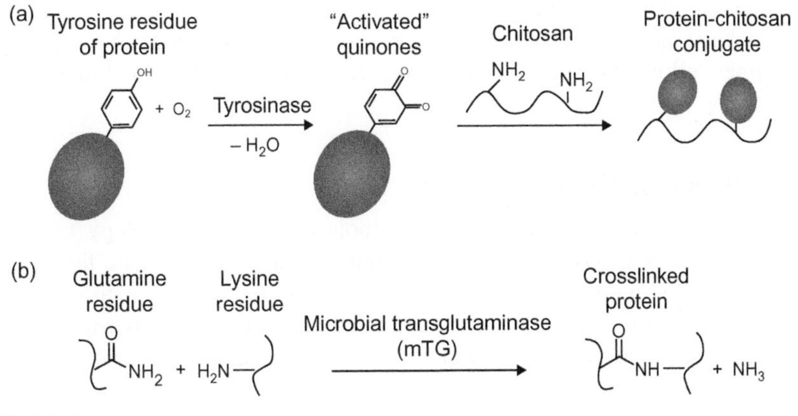

FIGURE 12.3

Two enzymatic reactions commonly used to build macromolecular structure. (a) Tyrosinase oxidizes phenols or the phenolic moieties of proteins (e.g., tyrosine residues) to generate o-quinones that can react with nucleophiles such as the primary amines of chitosan. (b) Microbial transglutaminase catalyzes the conjugation/crosslinking of proteins through glutamine and lysine residues.

to chitosan to generate a stimuli-responsive protein-chitosan conjugate that can be electrodeposited by the neutralization mechanism of Figure 12.2a [35,36].

The second enzyme we study is a microbial transglutaminase (mTG) that catalyzes the transamidation of lysine and glutamine residues of proteins to generate N-ε-(γ-glutamyl)lysine crosslinks (Figure 12.3b). The best-known mammalian transglutaminase is Factor XIIIa, which catalyzes the crosslinking of fibrin in the last stage of the blood coagulation cascade. The microbial transglutaminase (mTG) is Ca^{2+}-independent and catalyzes reactions with a broad range of proteins. This enzyme does not appear to have a strong sequence-specificity and appears specific for the glutamine but not lysine residues. Thus, mTG is receiving increasing attention as a means to generate crosslinked hydrogels, [37] form dimeric proteins [38,39] and macromolecular conjugates, [40,41] and site-selectively modify [42,43] and immobilize proteins [44].

Importantly, tyrosinase, mTG, and presumably other enzymes can only react with amino acid residues that are accessible. As a result of this constraint, many of the early studies with tyrosinase and mTG focused on reactions with open chain proteins such as gelatin [45,46]. The residues of globular proteins (e.g., enzymes and antibodies) are often inaccessible. As will be discussed in the following sections, a common strategy is to genetically engineer globular proteins with short fusion tails that provide accessible amino acid residues for enzymatic conjugation.

In summary, enzymes provide a simple, selective, and "biocompatible" means to generate macromolecular conjugates and stable crosslinked networks. The ability to engineer proteins with fusion tags of accessible residues may broaden enzymatic-assembly to make it generic: the protein's native structure no longer needs to have an accessible residue for reaction. In addition, the selectivity of enzymes allows orthogonal strategies to build structure or functionalize surfaces [47].

12.2.3 Biomolecular engineering

There is immense potential for using the tools of molecular biology to impart assembly functions to proteins. We cite a few examples to illustrate the breadth of opportunity to engineer proteins for self-assembly or for enzymatic-assembly. Histidine tags are routinely used in biotechnology to confer proteins with the ability to selectively and reversibly bind to metal-functionalized surfaces. Cysteine residues provide another self-assembly approach for proteins—in this case through reversible di-sulfide linkages. Proteins have also been engineered with longer amino acid sequences to permit self-assembly (e.g., using leucine zippers [48] or elastin-like polypeptides [49]). If known peptide sequences are inadequate, then discovery-based (e.g., directed evolution) approaches could be used to generate novel sequences for binding to specific surfaces [50]. Larger fusions could also be considered (e.g., with protein domains) to broaden the options for self-assembly even further. For instance, a protein G domain was engineered into an enzyme-fusion protein: binding of the protein G domain to an IgG antibody provided a versatile means to connect the enzyme-fusion with an antibody-based recognition/targeting moiety [51]. Importantly, the self-assembly approaches described above can, in principle, be extended from proteins to cells through surface display.

In addition to engineering proteins for self-assembly, several groups have engineered proteins with sequences that facilitate enzymatic-assembly. As mentioned earlier, it has often been observed that the amino acid residues of globular proteins are inaccessible, and fusion tags are required to provide the accessible residues for enzymatic-assembly. Several groups have reported such fusion tags to facilitate protein assembly through peroxidase [30,52] microbial transglutaminase [38] and sortase [53] enzymes.

12.3 Biofabrication of a multifunctional matrix

The example in Figure 12.4 illustrates the capability of biofabricating soft matter to have multiple functional capabilities [54]. In this example, we organized a hydrogel matrix based on the biocompatible protein gelatin, and this matrix was functionalized with both cellular and molecular activities. Gelatin is a well-known thermally responsive gel-forming polymer, yet no mechanisms are known for its direct electrodeposition. As illustrated in Figure 12.4a, we co-deposited gelatin using the pH-responsive small molecule hydrogelator 9-fluorenylmethoxycarbonyl-phenylalanine (Fmoc-Phe) that can be triggered to form a fibrous gel under slightly acidic conditions [55–57].

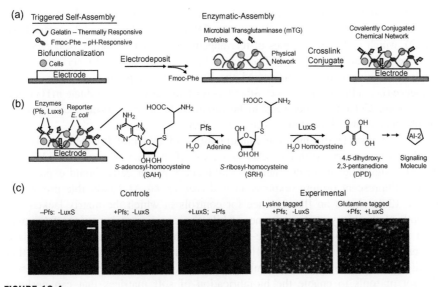

FIGURE 12.4

Fabrication and demonstration of the multiple functionalities of a gelatin-based soft matrix. (a) The small molecule hydrogelator (Fmoc-Phe) is used to electrodeposit gelatin and *E. coli* reporter cells. Microbial transglutaminase (mTG) is used to crosslink gelatin and conjugate two enzymes (Pfs and LuxS) to the matrix. (b) The pathway steps that lead to the formation of the AI-2 quorum sensing signaling molecule. (c) Experimental results show that both pathway enzymes are required to induce expression of fluorescent protein from the entrapped reporter cells.

Source: Reprinted from reference [54] by permission of John Wiley & Sons, Inc.

E. coli reporter cells were also incorporated into the warm (37°C) deposition solution and co-deposited with gelatin in the Fmoc-Phe matrix. After co-deposition, the film is cooled to room temperature to trigger gelatin's sol-gel transition. Interestingly, the subsequent incubation of this matrix under slightly basic conditions (pH ≥ 7.4) allows Fmoc-Phe to leach from the matrix, and thus this small molecule hydrogelator serves as a temporary fabrication aid to allow for the electrodeposition of gelatin.

The gelatin gel generated by this co-deposition is a physical (i.e., reversible) network that can be redissolved by mild heating. To "stabilize" this gelatin network, we enzymatically introduced covalent crosslinks by incubating the electrodeposited gelatin with microbial transglutaminase (mTG). This crosslinking mechanism is sufficiently mild that the entrapped cells retain viability and can proliferate [37].

In addition to catalyzing gelatin crosslinking, we used mTG to covalently conjugate two proteins to the gelatin matrix. These proteins catalyze sequential reactions in a short pathway for the biosynthesis of a bacterial quorum-sensing molecule autoinducer-2 (AI-2). As illustrated in Figure 12.4b, the first enzyme (designated Pfs) converts *S*-adenosyl-homocysteine (SAH) to *S*-ribosyl-homocysteine (SRH), while the second enzyme (designated LuxS) converts SRH to 4,5-dihydroxy-2,3-pentanedione (DPD), which undergoes a series of rearrangements to yield a family of compounds that are referred to as AI-2. Consistent with previous efforts to perform mTG-catalyzed conjugation, we engineered the Pfs and LuxS proteins to have a short sequence (i.e., a fusion tag) of accessible Lys (or Gln) residues to facilitate mTG catalysis. (Note: Gelatin did not need to be modified because its open chain structure enables its Lys and Gln residues to be accessible.)

To demonstrate the multiple functionalities of this biofabricated matrix, we incubated this matrix with the SAH precursor. If the enzymes convert SAH into the AI-2 signaling molecule, then the entrapped reporter cells should express the DsRed fluorescent protein (expression is driven by an AI-2-inducible promoter). Figure 12.4c shows no fluorescence for controls in which the matrix lacked one or both of the biosynthetic enzymes, while the experimental matrices fabricated with either lysine- or glutamine-tagged enzymes became fluorescent in the presence of the SAH precursor [54]. These results in Figure 12.4 demonstrate that electrodeposition, enzymatic-assembly, and biomolecular-engineering are a versatile set of tools to enable the biofabrication of soft matrices that can perform multiple, diverse functions.

12.4 Postfabrication biofunctionalization of devices

A major motivation for integrating biology with devices is to apply the power of electronics to the life and medical sciences. In many cases, microfluidic devices are envisioned either to permit lab-on-chip biosensing capabilities (e.g., for multiplexed analyses at the point-of-care) or to provide in vitro experimental tools to study complex biological phenomena (e.g., multicompartment drug metabolism [58]). A key challenge is assembling the labile biological components/systems at specific "addresses" within the microfluidic device. While various methods can be envisioned to assemble the biology, many biofunctionalization methods require compromises. For instance, if printing methods are used for assembly, the microfluidic system cannot be fully assembled until after biofunctionalization is complete. If assembly leads to a "permanent" structure (e.g., a nondegradable photocrosslinked polymer), then it may be difficult to wash the device for reuse. If simpler, less expensive devices are created for single-use applications, then it will be cost-prohibitive to have a device with the most powerful suite of electronic capabilities.

As illustrated in Figure 12.5, we aim to separate device fabrication from biofunctionalization by creating tools that allow the biological systems to be assembled on demand into previously fabricated devices [59]. We believe this approach will allow devices to be created using the most advanced microfabrication techniques without requiring compromises to accommodate biofunctionalization. On-demand biofunctionalization is enabled by electroaddressing, in which the biological

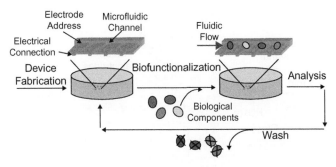

FIGURE 12.5

Postfabrication biofunctionalization allows devices to be fully fabricated without compromise. After fabrication, device-imposed electrical signals are employed to assemble the biological components for biofunctionalization.

Source: Adapted from reference [2] with permission from The Institute of Physics Publishing.

components are directed to assemble in response to device-imposed electrical signals (e.g., by electrodeposition). If the assembled structures are not permanent (e.g., biopolymeric hydrogels), then they can be removed after use, the channels can be washed, and the device can be reused. Thus, the ability to biofunctionalize the device postfabrication should limit the compromises required to integrate biology with the electronics. Further, the ability to reuse the device should allow cost-effective access to the power of electronics for collecting, processing, and transmitting data.

12.5 Test device

To illustrate the capabilities of postfabrication biofunctionalization, we created the simple test device illustrated in Figure 12.6. This device is fabricated from two glass slides, each patterned with 1-mm-wide parallel gold lines and positioned to be 1 mm apart to form the fluidic channel. The gold lines on these glass slides serve as the sidewall electrode addresses and their associated leads to an external power supply. The base and cover for the device are generated from polydimethylsiloxane (PDMS), which allows real-time optical and spectroscopic imaging of the deposited film, as illustrated in Figure 12.6. Initial studies with the device examined the mechanisms and controllability of the cathodic deposition of chitosan (Figure 12.2a) [14] and the anodic deposition of Ca^{2+}-alginate (Figure 12.2b) [17].

12.6 Electroaddressing a biosensing enzyme

It is often envisioned that lab-on-chip devices could provide platforms for rapid, automated, miniaturized, and remote analysis. Often this vision includes a

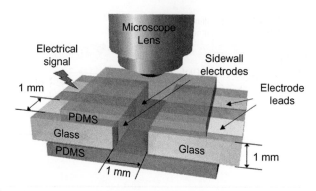

FIGURE 12.6

Schematic of test microfluidic device with built-in sidewall electrodes for biofunctionalization and with transparent cover for visual observation.

biological recognition element to confer selectively and the transduction of this recognition into a convenient, device-compatible output. A common model biosensing enzyme is glucose oxidase (GOx), which is often the biological recognition element used for home-use glucose measuring devices. GOx catalyzes the selective-oxidation of glucose and generates hydrogen peroxide that can be detected electrochemically to generate an output current that is quantitatively related to the glucose level.

$$Glucose + O_2 \rightarrow Gluconic\,Acid + H_2O_2$$
$$H_2O_2 \rightarrow O_2 + 2H^+ + 2e^-$$

We routinely use GOx as a model electrochemical biosensing enzyme to evaluate the capabilities of various electroaddressing options. A recently discovered electroaddressing option involves the anodic oxidation of chitosan by the putative mechanism of Figure 12.7a. According to this mechanism, Cl^- is anodically oxidized to form Cl_2, which then reacts with water to form the HOCl species. HOCl is a reactive chlorine species that can partially oxidize polysaccharides such as chitosan and in some cases generate aldehyde moieties. Aldehydes are convenient functional groups because they can undergo Schiff base-formation with primary amines. Both physical and chemical evidence support the putative mechanism of Figure 12.7a. Importantly, if aldehydes are generated by this mechanism, then Schiff base-formation with amines from another chitosan chain could generate a crosslinked chitosan network, while reaction with an amine from a protein could covalently conjugate the protein to chitosan. Thus, this anodic deposition mechanism offers the potential for a single-step approach to both deposit chitosan and activate it for the covalent conjugation of proteins [60].

We evaluated the potential of this anodic deposition mechanism for addressing GOx to an electrode address within the fluidic test device. For this test we mixed

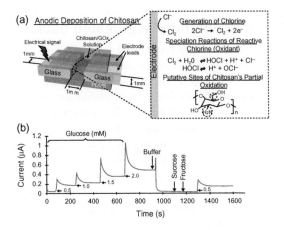

FIGURE 12.7

Programmable assembly of biosensing functionality at an electrode address of a fluidic device. (a) Schematic of the proposed anodic mechanism for the simultaneous deposition and conjugation of protein to chitosan. (b) Quantitative electrochemical biosensing using the model enzyme glucose oxidase (GOx).

Source: Adapted with permission from reference [60]. Copyright © 2012 American Chemical Society.

GOx (680 U/mL) into the deposition solution (1% chitosan dissolved in acetic acid with 0.15 M NaCl), added it to the fluidic channel, and applied an anodic potential to a sidewall electrode address (constant current of 4 A/m^2 for 90 s). After washing the deposit by flowing buffer through the channel, the electrode below the deposited gel was poised +0.6 V (vs a Ag/AgCl wire quasi-reference electrode) to detect H$_2$O$_2$, and the channel was filled with sugar solutions of varying concentrations. Figure 12.7b shows that each time the glucose level of the test solution was increased, the output current increased. Further observations from the experiment are (1) the steady-state output current was proportional to the glucose concentration; (2) minimal output currents were observed when the glucose was deleted from the test solution or when glucose was replaced by other sugars; and (3) the output currents were similar before and after washing the deposited gel with a detergent [60]. This latter observation supports the hypothesis that GOx is covalently grafted to the anodically deposited chitosan network.

From a broader perspective, the results in Figure 12.7 illustrate that electrode-position allows biological components to be rapidly assembled at electrode addresses in fully packaged fluidic devices by enlisting device-imposed electrical signals. As illustrated by the GOx example, assembly at an electrode address is particularly convenient when signal transduction involves electrochemical mechanisms. We envision that electroaddressing can be extended to multiple different biosensing addresses (for multiplexed analysis) and to other signal transduction modalities (e.g., optical [61] and mechanical [62]).

12.7 Creation of a spatially segregated bacterial biofilm

Bacterial biofilms are increasingly recognized for their roles in health (the microbiome), disease (drug-resistance), and technology (the biofouling of surfaces). Often these bacteria are entrapped within a self-generated extracellular polymeric matrix and are phenotypically different from planktonic cells [63]. Despite the importance of bacterial biofilms, there are few experimental tools available to assemble complex, stratified populations in vitro and to study their behavior over time. We are using the test device in Figure 12.6 as such an experimental tool and co-depositing bacteria within a Ca^{2+}-alginate matrix [64]. Importantly, alginate is a common component of the extracellular polymeric matrix generated by bacteria (e.g., by *Pseudomonas aeruginosa*), and thus we propose that co-deposition of bacteria within alginate provides an experimentally controllable model of a bacterial biofilm [65].

To illustrate the potential of electrodeposition for creating complex biofilms, we sequentially deposited three different populations of *E. coli* on a sidewall electrode address of the test device, as illustrated in Figure 12.8 [65]. Each population could be distinguished by the expression of a separate fluorescent protein as shown in Figure 12.8a. The initial deposition step was performed by filling the channel with a solution containing alginate (1%), $CaCO_3$ (0.25%), and the red-fluorescent-protein (RFP)-expressing *E. coli* (optical density of 6−8) and applying an anodic potential (4 A/m^2 for 2 minutes). After deposition, the channel

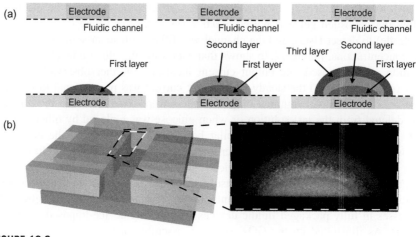

FIGURE 12.8

Multilayer cell assembly on individual electrodes of a fluidic device. (a) Schematic of the sequential deposition of *E. coli* cells that express red, green, and blue fluorescent proteins into a multilayer "biofilm." (b) Schematic and fluorescence photomicrograph of the multilayer.

Source: Adapted from reference [65] by permission of John Wiley & Sons, Inc.

was rinsed with 1 mM NaCl for 30 seconds. Next, the channel was filled with a second deposition solution—in this case with *E. coli* expressing green-fluorescent protein (GFP)—and deposition was performed at 4 A/m^2 for 4 minutes. After rinsing, the channel was filled with a third deposition solution containing an *E. coli* population expressing blue-fluorescent protein (BFP), and deposition was repeated—in this case for 6 minutes. The fluorescence photomicrograph in Figure 12.8b demonstrates that sequential co-deposition can generate a "biofilm" with stratified and segregated populations of bacteria. Subsequent studies demonstrated that the co-deposited cells recognize and respond to environmental stimuli (i.e., they can be induced to express genes in response to chemical inducers). Also, entrapped populations were observed to synthesize and respond to localized quorum sensing signaling molecules [65].

In summary, the results in Figure 12.8 illustrate the potential of biopolymer electroaddressing to assemble a complex biological system in a fluidic device. We envision that such devices could provide valuable in vitro experimental tools to enable the study of biological phenomena (e.g., quorum sensing) over time and length scales that are often difficult to achieve by alternative methods.

CONCLUSION

Over the Past half century, biotechnology and electronics each provided transformational capabilities, and we anticipate further transformations by the effective coupling of these technologies. We are pursuing the use of biology's fabrication approaches to build the bio-device interface, and we provide illustrative examples to suggest the potential for ever more powerful tools to analyze samples or to investigate biology. Yet, we believe that methods that enable the effective coupling of biology with microfabricated devices will have broader impacts [66]. For instance, the capabilities of synthetic biology could be accessed to allow the direct conversion of chemical energy into electricity [67] or the use of electrical energy to drive biosynthesis [68,69], thus allowing a fuller utilization of renewable resources to sustainably meet our needs for energy, chemicals, and materials. Further, the potential to "connect" devices to the body to allow two-way communication through mechanical, chemical, and electrical modalities would enable imaging to be more interactive, theranostics to be more personalized, and prosthetics to be "lifelike." Finally, it is possible to envision the use of devices that serve as manufacturing platforms to provide the precise spatiotemporal sequence of inputs (e.g., electrical or fluidic stimuli) to organize soft matter into multifunctional matrices or particles. In sum, biology provides useful materials and valuable lessons that we believe can be applied to integrate the capabilities of electronics with the versatility of biology!

Acknowledgment

The authors gratefully acknowledge financial support from the Robert W. Deutsch Foundation and the National Science Foundation (NSF; EFRI-0735987).

References

[1] Yi HM, Wu LQ, Bentley WE, Ghodssi R, Rubloff GW, et al. Biofabrication with chitosan. Biomacromolecules 2005a;6:2881–94.
[2] Liu Y, Kim E, Ghodssi R, Rubloff GW, Culver JN, et al. Biofabrication to build the biology-device interface. Biofabrication 2010;2: 022002.
[3] Koev ST, Dykstra PH, Luo X, Rubloff GW, Bentley WE, et al. Chitosan: an integrative biomaterial for lab-on-a-chip devices. Lab Chip 2010;10:3026–42.
[4] Cheng Y, Luo X, Payne GF, Rubloff GW. Biofabrication: programmable assembly of polysaccharide hydrogels in microfluidics as biocompatible scaffolds. J Mater Chem 2012a;22:7659–66.
[5] Braun E, Eichen Y, Sivan U, Ben-Yoseph G. DNA-templated assembly and electrode attachment of a conducting silver wire. Nature 1998;391:775–8.
[6] Nam KT, Kim DW, Yoo PJ, Chiang CY, Meethong N, et al. Virus-enabled synthesis and assembly of nanowires for lithium ion battery electrodes. Science 2006;312:885–8.
[7] Royston E, Ghosh A, Kofinas P, Harris MT, Culver JN. Self-assembly of virus-structured high surface area nanomaterials and their application as battery electrodes. Langmuir 2008;24:906–12.
[8] Wu LQ, Gadre AP, Yi HM, Kastantin MJ, Rubloff GW, et al. Voltage-dependent assembly of the polysaccharide chitosan onto an electrode surface. Langmuir 2002;18:8620–5.
[9] Redepenning J, Venkataraman G, Chen J, Stafford N. Electrochemical preparation of chitosan/hydroxyapatite composite coatings on titanium substrates. J Biomed Mater Res A 2003;66:411–6.
[10] Fernandes R, Wu LQ, Chen TH, Yi HM, Rubloff GW, et al. Electrochemically induced deposition of a polysaccharide hydrogel onto a patterned surface. Langmuir 2003;19:4058–62.
[11] Luo XL, Xu JJ, Du Y, Chen HY. A glucose biosensor based on chitosan-glucose oxidase-gold nanoparticles biocomposite formed by one-step electrodeposition. Anal Biochem 2004;334:284–9.
[12] Pang X, Zhitomirsky I. Electrodeposition of composite hydroxyapatite-chitosan films. Mater Chem Phys 2005;94:245–51.
[13] Zangmeister RA, Park JJ, Rubloff GW, Tarlov MJ. Electrochemical study of chitosan films deposited from solution at reducing potentials. Electrochim Acta 2006;51:5324–33.
[14] Cheng Y, Luo XL, Betz J, Buckhout-White S, Bekdash O, et al. In situ quantitative visualization and characterization of chitosan electrodeposition with paired sidewall electrodes. Soft Matter 2010;6:3177–83.

[15] Cheong M, Zhitomirsky I. Electrodeposition of alginic acid and composite films. Colloid Surface A 2008;328:73−8.

[16] Shi XW, Tsao C-Y, Yang X, Liu Y, Dykstra P, et al. Electroaddressing of cell populations by co-deposition with calcium alginate hydrogels. Adv Funct Mater 2009;19:2074−80.

[17] Cheng Y, Luo XL, Betz J, Payne GF, Bentley WE, et al. Mechanism of anodic electrodeposition of calcium alginate. Soft Matter 2011a;7:5677−84.

[18] Wu LQ, Lee K, Wang X, English DS, Losert W, et al. Chitosan-mediated and spatially selective electrodeposition of nanoscale particles. Langmuir 2005;21:3641−6.

[19] Ma R, Zhitomirsky I. Electrophoretic deposition of chitosan-albumin and alginate-albumin films. Surf Eng 2011;27:51−6.

[20] Luo XL, Xu JJ, Wang JL, Chen HY. Electrochemically deposited nanocomposite of chitosan and carbon nanotubes for biosensor application. Chem Commun 2005;2169−71.

[21] Yang XH, Kim E, Liu Y, Shi XW, Rubloff GW, et al. In-film bioprocessing and immunoanalysis with electroaddressable stimuli-responsive polysaccharides. Adv Funct Mater 2010;20:1645−52.

[22] Boccaccini AR, Keim S, Ma R, Li Y, Zhitomirsky I. Electrophoretic deposition of biomaterials. J R Soc Interface 2010;7:S581−613.

[23] Hao C, Ding L, Zhang X, Ju H. Biocompatible conductive architecture of carbon nanofiber-doped chitosan prepared with controllable electrodeposition for cytosensing. Anal Chem 2007;79:4442−7.

[24] Jiang T, Zhang Z, Zhou Y, Liu Y, Wang ZW, et al. Surface functionalization of titanium with chitosan/gelatin via electrophoretic deposition: characterization and cell behavior. Biomacromolecules 2010;11:1254−60.

[25] Wang Y, Pang X, Zhitomirsky I. Electrophoretic deposition of chiral polymers and composites. Colloid Surface B 2011b;87:505−9.

[26] Jin Z, Güven G, Bocharova V, Halámek J, Tokarev I, Minko S, Melman A, Mandler D, Katz E. Electrochemically controlled drug-mimicking protein release from iron-alginate thin-films associated with an electrode. ACS Appl Mater Interfaces 2011;4:466−75.

[27] Wang Y, Liu Y, Cheng Y, Kim E, Rubloff GW, et al. Coupling electrodeposition with layer-by-layer assembly to address proteins within microfluidic channels. Adv Mater 2011a;23:5817−21.

[28] Yang XH, Liu Y, Payne GF. Crosslinking lessons from biology: enlisting enzymes for macromolecular assembly. J Adhes 2009a;85:576−89.

[29] Teixeira LS, Feijen J, van Blitterswijk CA, Dijkstra PJ, Karperien M. Enzyme-catalyzed crosslinkable hydrogels: emerging strategies for tissue engineering. Biomaterials 2012;33:1281−90.

[30] Stayner RS, Min DJ, Kiser PF, Stewart RJ. Site-specific cross-linking of proteins through tyrosine hexahistidine tags. Bioconjug Chem 2005;16:1617−23.

[31] Chan L, Cross HF, She JK, Cavalli G, Martins HF, et al. Covalent attachment of proteins to solid supports and surfaces via Sortase-mediated ligation. PLoS One 2007;2: e1164.

[32] Bakota EL, Aulisa L, Galler KM, Hartgerink JD. Enzymatic cross-linking of a nanofibrous peptide hydrogel. Biomacromolecules 2011;12:82−7.

[33] Jus S, Stachel I, Fairhead M, Meyer M, Thony-Meyer L, et al. Enzymatic cross-linking of gelatine with laccase and tyrosinase. Biocatal Biotransform 2012;30:86−95.

[34] Yang Z, Liang G, Xu B. Enzymatic hydrogelation of small molecules. Acc Chem Res 2008;41:315—26.

[35] Chen TH, Small DA, Wu LQ, Rubloff GW, Ghodssi R, et al. Nature-inspired creation of protein-polysaccharide conjugate and its subsequent assembly onto a patterned surface. Langmuir 2003c;19:9382—6.

[36] Yi HM, Wu LQ, Ghodssi R, Rubloff GW, Payne GF, et al. Signal-directed sequential assembly of biomolecules on patterned surfaces. Langmuir 2005b;21:2104—7.

[37] Chen T, Small DA, McDermott MK, Bentley WE, Payne GF. Enzymatic methods for in situ cell entrapment and cell release. Biomacromolecules 2003;4:1558—63.

[38] Tanaka T, Kamiya N, Nagamune T. Peptidyl linkers for protein heterodimerization catalyzed by microbial transglutaminase. Bioconjug Chem 2004;15:491—7.

[39] Kamiya N, Tanaka T, Suzuki T, Takazawa T, Takeda S, et al. S-peptide as a potent peptidyl linker for protein cross-linking by microbial transglutaminase from *Streptomyces mobaraensis*. Bioconjug Chem 2003;14:351—7.

[40] Sato M, Furuike T, Sadamoto R, Fujitani N, Nakahara T, et al. Glycoinsulins: dendritic sialyloligosaccharide-displaying insulins showing a prolonged blood-sugar-lowering activity. J Am Chem Soc 2004a;126:14013—22.

[41] Sato M, Sadamoto R, Niikura K, Monde K, Kondo H, et al. Site-specific introduction of sialic acid into insulin. Angew Chem Int Ed 2004b;43:1516—20.

[42] Josten A, Meusel M, Spener F. Microbial transglutaminase-mediated synthesis of hapten-protein conjugates for immunoassays. Anal Biochem 1998;258:202—8.

[43] Villalonga R, Fernandez M, Fragoso A, Cao R, Di Pierro P, et al. Transglutaminase-catalyzed synthesis of trypsin-cyclodextrin conjugates: kinetics and stability properties. Biotechnol Bioeng 2003;81:732—7.

[44] Tominaga J, Kamiya N, Doi S, Ichinose H, Maruyama T, et al. Design of a specific peptide tag that affords covalent and site-specific enzyme immobilization catalyzed by microbial transglutaminase. Biomacromolecules 2005;6:2299—304.

[45] Chen T, Embree HD, Wu LQ, Payne GF. In vitro protein-polysaccharide conjugation: tyrosinase-catalyzed conjugation of gelatin and chitosan. Biopolymers 2002;64:292—302.

[46] Chen T, Embree HD, Brown EM, Taylor MM, Payne GF. Enzyme-catalyzed gel formation of gelatin and chitosan: potential for in situ applications. Biomaterials 2003;24:2831—41.

[47] Yang XH, Shi XW, Liu Y, Bentley WE, Payne GF. Orthogonal enzymatic reactions for the assembly of proteins at electrode addresses. Langmuir 2009b;25:338—44.

[48] Kostelny SA, Cole MS, Tso JY. Formation of a bispecific antibody by the use of leucine zippers. J Immunol 1992;148:1547—53.

[49] Hyun J, Lee WK, Nath N, Chilkoti A, Zauscher S. Capture and release of proteins on the nanoscale by stimuli-responsive elastin-like polypeptide "switches." J Am Chem Soc 2004;126:7330—5.

[50] Baneyx F, Schwartz DT. Selection and analysis of solid-binding peptides. Curr Opin Biotechnol 2007;18:312—7.

[51] Fernandes R, Luo X, Tsao CY, Payne GF, Ghodssi R, et al. Biological nanofactories facilitate spatially selective capture and manipulation of quorum sensing bacteria in a bioMEMS device. Lab Chip 2010;10:1128—34.

[52] Minamihata K, Goto M, Kamiya N. Protein heteroconjugation by the peroxidase-catalyzed tyrosine coupling reaction. Bioconjug Chem 2011;22:2332—8.

[53] Parthasarathy R, Subramanian S, Boder ET. Sortase A as a novel molecular "stapler" for sequence-specific protein conjugation. Bioconjug Chem 2007;18:469—76.

[54] Liu Y, Terrell JL, Tsao CY, Wu H-C, Javvaji VEK, et al. Biofabricating multi-functional soft matter with enzymes and stimuli-responsive materials. Adv Funct Mater 2012;22:3004—12.

[55] Liu Y, Cheng Y, Wu HC, Kim E, Ulijn RV, et al. Electroaddressing agarose using fmoc-phenylalanine as a temporary scaffold. Langmuir 2011a;27:7380—4.

[56] Liu Y, Kim E, Ulijn RV, Bentley WE, Payne GF. Reversible electroaddressing of self-assembling amino-acid conjugates. Adv Funct Mater 2011b;21:1575—80.

[57] Johnson EK, Adams DJ, Cameron PJ. Directed self-assembly of dipeptides to form ultrathin hydrogel membranes. J Am Chem Soc 2010;132:5130—6.

[58] Esch MB, King TL, Shuler ML. The role of body-on-a-chip devices in drug and toxicity studies. Annual Review Biomed Engg 2011;13:55—72.

[59] Park JJ, Luo X, Yi H, Valentine TM, Payne GF, et al. Chitosan-mediated in situ biomolecule assembly in completely packaged microfluidic devices. Lab Chip 2006;6:1315—21.

[60] Gray KM, Liba BD, Wang Y, Cheng Y, Rubloff GW, et al. Electrodeposition of a biopolymeric hydrogel: potential for one-step protein electroaddressing. Biomacromolecules 2012;13:1181—9.

[61] Liu Y, Gaskell KJ, Cheng Z, Yu LL, Payne GF. Chitosan-coated electrodes for bimodal sensing: selective post-electrode film reaction for spectroelectrochemical analysis. Langmuir 2008;24:7223—31.

[62] Koev ST, Powers MA, Yi H, Wu LQ, Bentley WE, et al. Mechano-transduction of DNA hybridization and dopamine oxidation through electrodeposited chitosan network. Lab Chip 2007;7:103—11.

[63] Flemming HC, Wingender J. The biofilm matrix. Nat Rev Microbiol 2010;8:623—33.

[64] Cheng Y, Luo XL, Tsao CY, Wu HC, Betz J, et al. Biocompatible multi-address 3D cell assembly in microfluidic devices using spatially programmable gel formation. Lab Chip 2011b;11:2316—8.

[65] Cheng Y, Tsao C-Y, Wu H-C, Luo X, Terrell JL, et al. Electroaddressing functionalized polysaccharides as model biofilms for interrogating cell signaling. Adv Funct Mater 2012b;22:519—28.

[66] Saliterman SS. Fundamentals of BioMEMS and medical microdevices. New York: Wiley Interscience; 2006.

[67] Tender LM. From mud to microbial electrode catalysts and conductive nanomaterials. Mrs Bull 2011;36:800—5.

[68] Ross DE, Flynn JM, Baron DB, Gralnick JA, Bond DR. Towards electrosynthesis in shewanella: energetics of reversing the mtr pathway for reductive metabolism. PLoS One 2011;6:e16649.

[69] Rabaey K, Rozendal RA. Microbial electrosynthesis—revisiting the electrical route for microbial production. Nat Rev Micro 2010;8:706—16.

Index

Note: Page numbers followed by "*f*" and "*t*" refer to figures and tables, respectively.

Printed and bound by CPI Group (UK) Ltd, Croydon, CR0 4YY

08/05/2025

01864838-0009